AF713064

Félix Guattari

Also by

Franco Berardi (Bifo)
Scrittura e movimento (Rome: Marsilio, 1970).
Chi ha ucciso Majakovski? (Milan: Squi/libri, 1977).
with Alain Guillerme and Pierre Rival, *L'ideologia francese: contro in 'nouveaux philosophes'* (Milan: Squi/libri, 1977).
Dell'innocenza: interpretazione del '77 (Bologna: Agalev, 1987).
Politiche della mutazione (Milan/Bologna: Synergon, 1991).
with Franco Bolelli, *60/90: dalla psichedelia alla telepatica* (Milan/Bologna: Synergon, 1992).
Mutazione e cyberpunk. Immaginario e tecnologia negli scenari di fine millennio (Milan: Costa & Nolan, 1994).
Come si cura il nazi (Rome: Castelvecchi, 1994).
Dell'innocenza: 1977, l'anno della premonizione (Verona: Ombre Corte, 1997).
La nefasta utopia di Potere Operaio (Rome: Castelvecchi, 1998).
La fabbrica dell 'infelicita': new economy e movimento del cognitariato (Rome: Derive-Approdi, 2001).
Il sapiente, il mercante, il guerriero. Dal rifiuto del lavoro all'emergere del cognitariato (Rome: Derive-Approdi, 2004).
Skizomedia. Trent'anni di mediattivismo (Rome: Derive-Approdi, 2006).

Giuseppina Mecchia
With Max Henninger and Timothy Murphy (eds) 'Italian Post-Workerist Thought', *SubStance* 112, Special Issue (2007).

Charles J. Stivale
The Two-Fold Thought of Deleuze and Guattari: Intersections and Animations (New York: Guilford Publications, 1998).
Disenchanting Les Bons Temps*: Identity and Authenticity in Cajun Music and Dance* (Durham, NC: Duke University Press, 2003).
Editor, *Gilles Deleuze: Key Concepts* (Chesham, UK: Acumen Publications/Montreal: McGill Queens University Press, 2005).
Gilles Deleuze's ABCs: The Folds of Friendship (Baltimore: Johns Hopkins University Press, 2008).

Félix Guattari

Thought, Friendship and Visionary Cartography

Franco Berardi (Bifo)

Translated and edited by
Giuseppina Mecchia and Charles J. Stivale

palgrave
macmillan

© Franco Berardi (Bifo) 2008
Translation © Giuseppina Mecchia and Charles J. Stivale 2008

All rights reserved. No reproduction, copy or transmission of this publication may be made without written permission.

No portion of this publication may be reproduced, copied or transmitted save with written permission or in accordance with the provisions of the Copyright, Designs and Patents Act 1988, or under the terms of any licence permitting limited copying issued by the Copyright Licensing Agency, Saffron House, 6-10 Kirby Street, London EC1N 8TS.

Any person who does any unauthorized act in relation to this publication may be liable to criminal prosecution and civil claims for damages.

The author has asserted his right to be identified as the author of this work in accordance with the Copyright, Designs and Patents Act 1988.

First published 2008 by
PALGRAVE MACMILLAN

Palgrave Macmillan in the UK is an imprint of Macmillan Publishers Limited, registered in England, company number 785998, of Houndmills, Basingstoke, Hampshire RG21 6XS.

Palgrave Macmillan in the US is a division of St Martin's Press LLC, 175 Fifth Avenue, New York, NY 10010.

Palgrave Macmillan is the global academic imprint of the above companies and has companies and representatives throughout the world.

Palgrave® and Macmillan® are registered trademarks in the United States, the United Kingdom, Europe and other countries.

ISBN-13: 978–0–230–22119–2 hardback
ISBN-10: 0–230–22119–X hardback

This book is printed on paper suitable for recycling and made from fully managed and sustained forest sources. Logging, pulping and manufacturing processes are expected to conform to the environmental regulations of the country of origin.

A catalogue record for this book is available from the British Library.

Library of Congress Cataloging-in-Publication Data

Berardi, Franco.
Felix Guattari : thought, friendship, and visionary cartography / Franco Berardi (Bifo); translated and edited by Giuseppina Mecchia and Charles J. Stivale.
p. cm.
Includes bibliographical references.
ISBN 13: 978–0–230–22119–2 (hardback)
ISBN 10: 0–230–22119–X (hardback)
1. Guattari, Flix, 1930–1992. I. Mecchia, Giuseppina, 1962– II. Stivale, Charles J. III, Title.

B2430.G784B4713 2008
194–dc22 2008020595

10 9 8 7 6 5 4 3 2 1
17 16 15 14 13 12 11 10 09 08

Transferred to Digital Printing 2011

Contents

Preface: Thought, Friendship and Visionary Cartography

Giuseppina Mecchia and Charles J. Stivale[1]

In one of Gilles Deleuze's final texts, a brief remembrance from August 1992, entitled simply 'For Félix', the eminent philosopher and writing partner concludes that Guattari's work 'is waiting to be discovered or rediscovered. That is one of the best ways to keep Félix alive'. Deleuze has already insisted that he refers not to their collaborative work, but precisely to Guattari's writing on his own. So, Deleuze closes his statement by reflecting: 'Perhaps the most painful aspects of remembering a dead friend are the gestures and glances that still reach us, that still come to us long after he is gone. Félix's work gives new substance to these gestures and glances, like a new object capable of transmitting their power' (2006, 383). It is fitting, then, that Franco Berardi, known familiarly as Bifo, frames his reflections on Guattari's work in terms of the power in the gestures and glances that constitute their enduring friendship.

In like fashion, the project to prepare an edition of these biographical reflections, written originally in Italian, is inspired by our friendship with Bifo and other readers of Guattari. Although the works of Gilles Deleuze have been translated and have generated an entire market of critical works, anthologies and guides to his thought, Félix Guattari has not achieved the same public exposure, especially given that very few writers have undertaken a serious engagement with his thought.[2] Besides the importance of making Guattari's works better known, we are also inspired by the experience of Guattari's practice lived by Bifo and expressed in his biography. As he explains in the introductory chapter, his book stems first from his acquaintance with Guattari's writings and political engagement in the context of Bifo's own activism in Italian autonomist politics in the 1970s and then his exile from Italy and direct collaboration with Guattari in Paris from 1977 onward. In light of the importance that Guattari placed on the political activities in Italy during the 1970s and 1980s, we propose to bring forward the way that Bifo illuminates the Italian political experience in relation to Guattari's life and thought.

Bifo's original work, published in 2001, is entitled *Félix. Narrazione dell'incontro con il pensiero di Guattari, cartografia visionaria del tempo che viene*: it translates as 'Félix. Tale of the encounter with Guattari's thought, a cartographic vision of the future'. To the translated text of this encounter, we add a succinct, critical introduction and a substantial interview with Bifo conducted during the summer of 2005. This biography gains distinction from its keen insight into Guattari's political practice and from a precise understanding of how this practice relates to the theoretical and conceptual aspects of Guattari's writings, alone and with Gilles Deleuze. One reason why key works from Guattari's corpus remain untranslated is their density and general inaccessibility due to a highly abstract theoretical language. Thanks to an approach that is at once personal and extremely well informed about the origins and development of Guattari's thought, Bifo's biography is quite clear and fairly devoid of the abstract language that characterizes Guattari's works.

Our approach in this Preface is two-fold: in a first section, we turn toward the historical perspective on Bifo's encounter with Guattari, in order to consider not only their shared political commitment through militant activism and a common understanding of late capitalism, but also the significance of the generational discrepancy between the two, mitigated somewhat by national differences. This historical perspective leads us to consider the concepts they shared, especially within the context of Bifo's work on Italian post-workerist politics. In the second section, we consider how Guattari and Deleuze together are re-constructed and inserted into this biography. That is, we have noted in reading *Félix* a conceptual level in the encounter of the two friends that is distinct from merely personal memories. One way of exploring this distinction from both Bifo's and Guattari's perspectives is through a review of *Félix* in light of the publication of Guattari's *The Anti-Oedipus Papers*, the collected notes written mostly by Guattari in preparing *Anti-Oedipus*. In light of these two concise and conjoined presentations, we hope to reveal facets of the complex web of interrelations that constitute the friendship between Guattari and Bifo.

I How could 'Bifo' meet 'Félix'?

It was the late Aldo Moro, the Italian statesman kidnapped and ultimately killed by the Red Brigades in 1978, who coined the paradoxical expression 'parallel convergences', in order to describe the trajectory of the Italian Communist Party and the Christian Democrats, the two forces that he was trying to bring together through his efforts toward reaching a 'historical compromise' between what had been the two strongest political actors in postwar Italy. It might not be inappropriate to start describing the improbable encounter between Franco Berardi 'Bifo' and Félix Guattari as a 'parallel convergence'. For, while there are perfectly solid historical circumstances that made that encounter possible, it is nonetheless in the infinite virtuality of creation that those circumstances became determinant and conceptually productive. And it is at infinity that parallels meet.

Félix Guattari was born in Villeneuve-les-Sablons in 1930, Bifo in Bologna in 1949. This generational and geographical distance was to be mediated in multiple ways. Some of these mediations may have been more imaginary than real, but nonetheless effective in drawing the two men together. For instance, when asked why he felt comfortable in writing a book on Félix Guattari, while he would not have done the same thing for Deleuze, Bifo said: 'Basically, I do not consider myself to be sufficiently a philosopher in order to address Deleuze in a direct way. While I seem to be just dirty enough from the disciplinary point of view ... to be able to approach [Guattari]' (Interview with 'Bifo', page 155 below). From a purely documentary point of view, that is simply not true, since Franco Berardi actually obtained a Laurea in Philosophy, specializing in aesthetics under the guidance of Italian phenomenologist Luciano Anceschi. Moreover, the characterization of Guattari as something 'other' than a philosopher might be more myth than reality, since if having a degree in philosophy doesn't make one a philosopher, then not having it should not necessarily prevent you from being a philosopher either.

In fact, we should state once and for all that the field of philosophy as production of concepts was and still is as much a common ground for Bifo and Guattari as is the messier, more controversial scene of political engagement. If some critics fail to recognize this commonality, it is due, in part at least, to the failure of memory associated with Félix Guattari and especially to the role that he played in Bifo's life. This role was such that the philosophical 'seriousness' both of Bifo and of Guattari tends to take second place in the usual characterizations of them.

Similarily, the vicissitudes of Bifo's political and personal trajectory are so compelling that they risk diverting our attention from the impressive coherence of his life's work. Fortunately, in the last few years, his writings have been attracting ever increasing interest in the Anglophone world, while in Italy, they remain somehow marginal and seem unable to attract serious academic attention. And this difficulty might be a true point of affinity with Guattari who himself, while always a rigorous, complex and stunningly fertile thinker, kept a busy political and personal agenda that might have made him appear a much more volatile thinker than he really was. The openness of Bifo and Guattari's multiple engagements was indeed what made their encounter possible – it would be difficult to imagine Bifo actually becoming a friend of the more reserved Gilles Deleuze – but we should not forget that this happened when their intellectual and political development had already made it possible for them to recognize each other as philosophical and political interlocutors.

Bifo was 27 when he first met Guattari in Paris during the summer of 1977 as the latter gave him refuge during his judicial travails. As Bifo himself says: 'I was already sufficiently trained to feel him as someone I understood perfectly' (Interview, 155). Trained, one should add, in the different but intimately related fields of conceptual and political creation. This double training was also at the very core of Guattari's own life and work.

As political activists, both Bifo and Guattari started their militant practices in their teens as members of Communist youth groups, and both left the Communist Party very early and contributed to the formation and the functioning of less hierarchical, authoritarian and politically compromised leftist groups. Bifo, in particular, became one of the early members of *Potere Operaio* (Workerist Power), together with figures such as Antonio Negri, Mario Tronti and Raniero Panzieri. This movement had a theoretical and practical importance within the tradition of Marxist thought which is only recently being appreciated by a large international audience.[3] At the beginning, following Gramsci's example, the movement focused on industrial workers, and it is against this bias that Bifo would later wage one of his most interesting critiques of *Potere Operaio*. While the group itself had a relatively short life (1969–72), its reflection on the autonomous power of subjugated groups had a long-lasting influence on Italian leftist political practices.[4] In particular, the different groups active around *Autonomia Operaia*, which was founded in 1973 and saw in Bifo one of its most influential members, sanctioned the divorce of revolutionary thought and practice from the sclerotic Italian Communist Party, and started paying very close attention to the changing nature of production and subjugation in late capitalist societies.

It is in this context that Bifo's contribution was to be essential: recognizing that the capitalist domination of media technologies was financially overwhelming but ontologically less than real, Bifo started focusing on the creation of autonomous sources of information, cultural production and affective participation in society. In 1974 Bifo published *Scrittura e Movimento*, an assessment of the cultural implications of the workers' struggles of the early 1970s. Written in essay form, this book takes on many of the themes staged narratively in *Vogliamo tutto* (1971) by fellow 'post-workerist' Nanni Balestrini who would also have to exile himself to Paris in the early 1980s. And it is also during these all-important years that Bifo read the 1974 Italian translation of *Anti-Oedipus*, where in fact Deleuze and Guattari articulate a similar critique of the primacy of authoritaritarian structures, both in psychological and politicals terms. Together with the work of Foucault on disciplinary societies and Baudrillard's research on the simulacra, this strand of French post-structuralist thought was most influential on the Italian autonomous movement in general, particularly in the media and culturally oriented university town of Bologna and on Bifo himself.[5] In 1975, Bifo and his Bolognese pals created the publishing house Squilibri, which produced and distributed, among other projects, the typewritten political and cultural magazine *A/traverso*, and Bifo's first novel, also typewritten (and still untranslated), *Chi ha ucciso Majakovski?* (*Who killed Mayakovsky?*). The creation of Radio Alice was perhaps the most enduring contribution made by Bifo to rethinking the conditions of cultural production in late-capitalist societies. This era was later to be called 'cognitive capitalism', and Bifo was among the first to recognize that the field of struggle between capitalist domination and autonomous subjective constitution is nothing less than the mind.

These activities and concerns help us understand why, when Bifo found himself in Paris fleeing from an arrest warrant in Italy, he was fully ready, politically and intellectually, to meet Félix and to engage in an encounter that was productive on all levels. Besides Marxism and various Marxist heresies, phenomenology, the formation of subjectivities and the construction of machinic assemblages oriented towards the liberation of the unconscious, the two activists also shared an interest in the thought of Gregory Bateson and were engaged in creating what Bateson had called 'an ecology of the mind'. During his stay in France, Bifo was also active on the Parisian intellectual scene, contributing to the rebuttal of the *Nouveaux Philosophes* – a philosophical and political project in which Guattari took a keen interest – with the untranslated publication in Italian of *L'Ideologia francese: contro i nuovi filosofi* (*French Ideology: Against the New Philosophers*), that he co-authored with Pierre Rival and Alain Guillerme (1977).

It is in fact in the late 1970s and early 1980s that the Marxist Left was being attacked in France on the wave of the supposed failure of May '68, but neither Bifo nor Guattari stopped their conceptual work, even while their more militant political activism was being challenged in myriad ways. Only during the 1980s did the generational difference between the two friends play an important role with regard to their individual choices as militants: Guattari seemed to keep a stricter communist revolutionary rhetoric and engagement during his collaboration with Antonio Negri, who arrived in Paris in 1983 and authored a book with Guattari in 1985, *Les Nouveaux espaces de liberté* (*Communists Like Us*) (Guattari and Negri 1990). At the same time, Guattari became increasingly involved with the French Green movements, even accepting to run, however unsuccessfully, as a candidate for one of their political formations. While always remaining quite close to Guattari, Bifo did not share his friend's activism during those years since, as we will see later, he was convinced that one must accept the inevitable 'depression' resulting from having lost at least some of one's personal and political struggles.

The two, however, continued to share the same fundamental interest: the affective, ecological and political consequences of what Guattari called the post-mediatic era or, alternatively and depending on the context, Integrated World Capitalism. The publication of *The Three Ecologies* and of *Chaosmosis* bears witness as much to Guattari's continuing reflection as to his activist support of mediatic initiatives such as participation in the project of Radio Tomate or his involvement in the Alter 36-15 Minitel initiative.[6] The influence of Bifo and his other Italian 'comrades' remained therefore of vital importance for Guattari's political and conceptual development during the 1980s and up to his death in 1992. And it is undoubtedly through Félix Guattari's enduring influence that the mediation between Italian autonomist and 'post-workerist' theories and practices and the Anglophone world was made possible, a filiation process that is now attracting a considerable critical interest.

It is therefore quite comprehensible that one of the few books on Félix Guattari is written by an Italian political activist and social critic. Moreover, and of even greater importance, the 'parallel convergences' that brought them together have not ceased to operate after Guattari's death. During the 1990s, Bifo continued to publish on many of the themes he had explored with his older friend, and to take part in media-oriented political practices. The continued timeliness of his intellectual engagement is making Bifo one of the most interesting thinkers of our time. Persistence might in fact be what is needed the most in the intellectual arena: it is precisely because the virtuality of the future never closes up that Guattari said in a late interview, 'Philosophy is essential to human existence' (2002, 19). And philosophers such as Bifo and Guattari are, unfortunately, few and far between.

II Friendship and community

One is struck in reading *Félix* by the profound sympathy and willingness that Bifo exercises in facing the truth about Guattari and their shared era, but it is a truth that required a deep reflection in order to state things as Bifo truly understood them, not as some might have liked them to be. Bifo concludes the autobiographical introductory chapter by expressing his desire to 'reconstruct the rhythmic map of Félix-thought', and then describes an even greater ambition, to employ that map to 'cause harmony to resonate among the chords, the refrains and the dissonances in the contemporary planetary rhapsody'. Given such a lofty (some would say utopian) goal, he then proceeds in a decidedly curious fashion in starting Part I by raising the question of depression. As we say in note 1 to Chapter 2, the translation we chose for the chapter title, 'The Happy Depression' (for 'La depressione Felix') plays with Guattari's name and its literal meaning. This choice of real starting point is, for Bifo, a necessary approach for understanding Félix-thought and the 'Félix-machine', not just the expansion of provisional community but also the possibility of its dissolution. Given that the book's final chapter evokes this relationship marvellously in the title, 'the provisional eternity of friendship', we can understand that Bifo wishes to frame the biography overall in relation to the fundamental fragility of existence in which friendship itself has an often crucial, yet ephemeral status.

So Bifo raises the little-discussed question of depression, or rather, little discussed in the context, on the one hand, of Guattari and his political and intellectual commitments and, on the other hand, of the theoretical concepts of desire and the politics of desiring community and activism. And little discussed philosophically, he argues: 'We have foreclosed [the experience of depression] and made it shameful.' Things fall apart, Bifo would say, and so he wants to speak to his friends and to Guattari's and consider frankly the consequences of political action in both of their lives, and how that action was related to a process of dissolution, also known as depression. As Bifo argues in the interview published at the end of this book, Guattari was unable to assess his aging process as well as 'his feminine side and his depression' since he lived his political and philosophical engagement in a heroic, 'real men' mode of thinking: 'There was the prevalence of the sense of historical presence, of historical duty, of the dialectic, which made it impossible for him to grasp the cognitive element that is present within depression' (Interview, 159).

Bifo is also not afraid to confront the intersection between Guattari's socio-political and philosophical engagement and the personal, since 'this story of [Félix's] depression ... we called it Josephine', that is, Guattari's life partner from 1986 onward. As Bifo admits in the interview, 'the Guattarian network [of friends] did not want to talk about her' (Interview, 160); an isolation occurred, then a break with friends, and Bifo concludes, 'so even in this, the relationship with femininity and with aging, with depression, became a crisis factor within the political community' (160). Yet Bifo's perspective on the importance of depression remains grounded in the philosophical and political, because depression has a deep core, he maintains, in the 'collapse of modern hope', as one realizes that his or her 'desire no longer has any place in the real' (160). However, this dire situation does not justify blinding oneself to these circumstances since such blindness results in an 'impotence of political will' – 'I am not saying that depression has a political origin, nor do I want to forget Josephine. I just want to say that depression is born out of the community's immediacy.'

This conjuncture with which Bifo opens his reflection inspired by a friendship with Guattari also points toward the important collaboration of Guattari with Deleuze. As Bifo understands their work, notably in *Anti-Oedipus*, 'the concept of desire [as the productive power of the unconscious] is linked to a youthful utopia deriving from Romanticism, a utopia which we should not disavow but cannot worship either'. Hence, Bifo argues for conceptualizing both sides of desire, as both productivity and as 'tension [that] is destined to fall, to die down'. This mature perspective is why Bifo comprehends the Deleuze-Guattari collaboration in its totality, contrasting the initial collaboration with the final one by evoking their words on friendship at the start of *What Is Philosophy?*, and he sees in this latter work the 'elaboration of a senile utopia'. On the basis of just such a 'sharing of experience' and 'understanding of its hallucinatory and thus impermanent character', he says, we might envision clearly 'the dissolution of the dependency and the attachment translating desire into depression' – in short, 'the way to nirvana' – 'the condition for being able to live desire while transcending it at the same time.'

These perspectives, taken solely from Chapter 2 (and from related material in the interview), draw attention to the manner in which Bifo inserts the Deleuze-Guattari conceptual relationship into his own narrative and reflections on Guattari. In the rest of Part I (chapters 3, 4 and 5), Bifo addresses different facets of the psycho-political dynamics on a global scale, about which Guattari blazed a trail in his different writings. Then, in Part II, especially in chapters 7 (on the rhizomatic machine), 8 (on *Anti-Oedipus*), and 9 (on *Kafka* and *A Thousand Plateaus*), Bifo explores different conceptual details that arose in the Deleuze-Guattari collaboration. But we believe it is important to keep in mind the opening chapter's comprehensive view of youth and maturity, integration and dissolution, hope and tension.

Bifo's approach stands in sharp contrast to a more recent publication, the working notes from the *Anti-Oedipus* collaboration, *The Anti-Oedipus Papers*. In Stéphane Nadaud's excellent introduction, humorously but quite appropriately entitled 'Love Story between an Orchid and a Wasp', he reflects on how the two-who-would-be-a-crowd (that is, Deleuze and Guattari) leave traces of each other in the drafts, notes and journal entries that constitute the volume. One pertinent question about this collaboration is: 'Is this what the collective aspect of enunciation amounts to, identifying something of Deleuze in Guattari and something of Guattari in Deleuze? Is it that simple?' (Guattari, *The Anti-Oedipus Papers*, 12). The common assumption about this collaboration had been that 'Guattari needed Deleuze in order to write. All the more so as Guattari made no bone [sic] about the fact that he certainly did' (12). Nadaud understands their collaboration somewhat differently from Bifo since Nadaud places it under the aegis of the concept of assemblage. For each of the writers generated texts that he sent to the other between their meetings, so *The Anti-Oedipus Papers* consists of letters by Guattari, notes on his readings, theoretical writing, and even his personal journal entries, all transmitted to Deleuze, with Fanny Deleuze serving quite crucially as frequent intermediary and also as editor, judging from many personal notes from Guattari to her. The assemblage, of course, did include Deleuze's contributions as well, since he evidently read closely everything and annotated much of what Guattari transmitted.

One gets a sense in *The Anti-Oedipus Papers* that after the two writers jammed and riffed, as it were, between one another, it fell to Deleuze to finalize the text and manuscript of *Anti-Oedipus*, at the risk of losing his identity in the process, according to Guattari, who understood by the end how much the process cost Deleuze. In preparing his own 'final' versions to deliver to Deleuze, Guattari consulted many specialists, for example for Chapter 3 of *Anti-Oedipus*, or availed the text of his own expertise in psychoanalysis, especially the perspectives on

Lacan. In any case, this dual intersection of relatively merged identities explains the image of the wasp and the orchid, as Nadaud says, a duality in order 'to conceive production as an assemblage of differences' (*The Anti-Oedipus Papers*, 20). In other words, they connected to each other rhizomatically, with their 'thought and writing ... flying off in every direction – even if in different ways (in a disordered, flowing way for Guattari; and a conceptualized, organized way for Deleuze; in multiple practices for Guattari; and with the solitude of an academic researcher for Deleuze)' (21). In contrast to Bifo's fairly macroscopic view of the Deleuze-Guattari collaboration, *The Anti-Oedipus Papers* is more microscopic, and indeed offers what is commonly termed 'too much information', except for the most ardent schizoanalyst. However, just as Bifo provides in his idiosyncratic manner an overview of the arc from *Anti-Oedipus* to *What Is Philosophy?*, *The Anti-Oedipus Papers* provides insight regarding the genesis of different concepts appearing subsequently in *Kafka. For a Minor Literature* and *A Thousand Plateaus* as well as in Guattari's own works.

The Anti-Oedipus Papers constitutes a record of a brief and intense moment in time, one that presumably represented at once a peak and a new phase in the reflection both of Deleuze and Guattari. Bifo came to know Guattari shortly thereafter, while Guattari was in the midst of developing *A Thousand Plateaus* and engaging in all manner of parallel political activities, including support for the Free Radio movement. So Bifo observed Guattari's work, close up and then from Italy, as it moved from the peak period of the late 1970s into the 'winter years' of the 1980s. Following Bifo's narrative, in contrast to the effervescence and intensity of the post-1968 collaboration with Deleuze, the final decade of Guattari's life was weighted down with the burdens of the zeitgeist, the expectation of his willing participation, and the demands of the aforementioned personal relationship.

Bifo's very personal reflection about his friend thus follows a specific trajectory from start to finish. As we indicated above, the second chapter, on happy depression, emphasized the ephemeral nature of friendship, and in the final chapter, that nature is taken up again, but in a different way. In closing, Bifo conjoins the key elements that sum up his understanding of the works of Félix Guattari, alone and in collaboration with Gilles Deleuze. He does so by creating a link between concepts produced through that collaboration, notably the Body without Organs (that Bifo considers, in his own terms, as a tantric egg), and the final works of these authors, Guattari's *Chaosmosis* and the collaborative *What Is Philosophy?*. As we have seen, the collaborative relationship with Gilles Deleuze was but one of several eminently productive incarnations of the conception of friendship espoused by Guattari during his lifetime. We are convinced that if being 'friends of thought' is, in fact, the hallmark of the philosophical project in general, then it must also be true that philosophers are friends with each other, independent from the fact that they are dead or alive. The creation of a philosophical portrait is, of course, the only true reason for the existence of the history of philosophy, or at least the type practised by the authors featured in this small, but truly philosophical book. And so, with the final return to the ephemeral nature of friendship, Bifo proposes to show, or at least to suggest, how the elements of irony, play, language, history and singularity intersect with the fundamental friendship at the core of Guattari's thought. These elements do not contradict, but rather complement and extend the possibility of happy depression, yet allow friendship to exist beyond distress as active affirmation of the project of philosophy.

1
Introduction: Cartographies in Becoming

Ever since Félix Guattari died in 1992, I have been promising myself that I would write this book.

But the book would never be finished, because rhizomatic thought is the cartography of landscapes yet to come, and so the landscapes in which this book proliferates have not ceased dispersing themselves before my eyes, with every passing day, faster than any light-speed writing.

The development of the telecommunication network, biomechanical proliferation, the Genome Project, the constitution of a bioinformational paradigm, all are successive manifestations of this becoming rhizome of the world, of which Félix projected the earliest maps.

In the meantime, the thought of Félix Guattari and Gilles Deleuze has gained a truly vast audience, especially over the Internet where a form of collective enunciation called the Web creates links and continues to proliferate.

With the insurrection in Seattle, 30 November 1999, this linkage revealed itself as a planetary political force. Collective agents of rhizomatic enunciation and insurrectional action are the same thing.[1]

Web linkage has truly set in motion a process through which both the thought of Deleuze-Guattari and the bibliography that it nourishes continue to proliferate, defeating any possibility of keeping pace with such proliferation.

Especially in the Anglo-American sphere, new books and reviews continue to appear on the themes that rhizomatic thought has brought to philosophical, psychoanalytic, political and aesthetic audiences.

The field of philosophical and political thought, the field of psychoanalysis, but also the field of biotechnology and cyberthinking are imbued with the principal concepts that Deleuze and Guattari's neological machine has constructed.

In the trilogy from *Anti-Oedipus* to *A Thousand Plateaus*, and then to *What Is Philosophy?*, an extraordinary intellectual adventure unfolds that probably concludes the arc of twentieth-century thought and transmits its vital energy into the thought of the next century.

****

I don't intend to summarize the contemporary success of Deleuze-Guattari thought. I simply want to tell my story, my encounter with this thought, and the perspectives that I see derived from it.

My encounter with Félix Guattari occurred at different moments. In 1974 when I was doing military service in punishment barracks in the south of Italy, I had decided to act like I was nuts so I could be sent back home.

A French friend had spoken to me about a psychoanalyst who was trying to see the world from the schizo's point of view instead of that of psychiatry, and so I bought one of his books, the only one that had come out in Italy.

It was called *Una tomba per Edipo* (*Psychanalyse et transversalité*) (1972).

One night in June, I became involved in a little act of craziness, refusing to abandon my turn at guard duty, and arguing that I would stay there until I ran out of strength. They sequestered me in the neuropsychiatric hospital in Naples, and after 10 days in observation, the medical officer asked to see me.

He asked me what was wrong.

I really didn't tell him anything, that everything was great, except that when I saw an automobile's licence plate, the numbers were then stamped into my brain, where they then went through all sorts of recombinations until I ended up with a headache.

The medical officer (named Moretti) looked at me for a moment with interest, and then said that if I had learned the lesson, I had learned it well. And he sent me home with a diagnosis of cenestropathic neurosis.

So in my sick mind was imbedded the idea that Félix had saved me from the barracks. You know, the flag-raising at 6.30am and all that running back and forth.

Then I read *Anti-Oedipus* in March 1976. At that time, I was in jail, in a cell in San Giovanni in Monte (a beautiful prison that was a convent in the fifteenth century and today houses the History Department of the University of Bologna). For all its great beauty, the prison depressed me, especially because they had accused me of having placed a bomb in the headquarters of the Christian Democrats, and I knew nothing about the deed. My friend Riccardo, who left then for a faraway destination and from time to time reappeared with a new Vietnamese or Californian wife, sent me a copy of *Anti-Oedipus* in jail. It was within this map of existential and theoretical wandering that I lost myself that year. Proliferating and losing oneself, this was the sense of collective enterprise that the movement was attempting in Italy.

In Bologna I was involved with a few friends in publishing *A/traverso*, a journal that had begun with the headline: *A small group in multiplication.*[2] The idea of contagion, of viral proliferation, was implicit in this formula presented as a model of organization (political? post-political? it hardly matters). And the idea that social processes, political and cultural transformations are contagions, proliferations of viruses that spread out in the social body and produce mutations – here is an idea that emerged from Félix's molecular vision. One of the contact points between rhizomatic thought and philosophical inspiration is William Burroughs, who spoke about language as a virus.

I met Félix in person only in June 1977.

A bizarre insurrection took place that year in Bologna that was more inspired by Dadaism and *Anti-Oedipus* than by political revolutionary manuals.

At a certain point, things went badly for me. I had spoken at several meetings, and had published some leaflets and newspapers. I often went to Rome where I met others in the Autonomist movement, and so a judge asserted that he had all the proof he needed to accuse me of fomenting class hate and the like.

In the meantime, terror had been unleashed in the cities. A kid killed a cop, everyday clashes in the urban centres, three hundred students arrested, young workers, even housewives who found themselves in the battle by chance. For a few days, I remained hidden in the city, staying at the home of a few friends, and then I hit the road to go abroad. Naturally to Paris. In June, I decided to call Félix. I don't remember the first meeting with him. I only know that he was suddenly what he has always has been since: a generous, innocent and ingenious friend.[3]

At the start of July they arrested me. The Italian judge who had it in for me came to Paris and convinced the local police that I was dangerous, and members of the anti-gang squad came to capture me while I was going to eat at a friend's house.

Shit, the *dépôt* of the La Santé prison is a fetid spot. About sixty of us were piled in a small room, while it was pouring rain outside, and you had to piss in the corner while waiting for something to happen.

I stayed there two days. Then I was transported to the Fresnes prison, which was already high-tech. You stayed alone in a cell, the walls were made of metal, and in the courtyard we had to walk in a single file. I actually missed the conventional prison in Bologna.

But it only lasted a week. Félix had got in touch with my friends, had activated the channels of communication among Parisian intellectuals, and had, in short, created the conditions for getting me released.

The judges had to recognize that the Italian tribunal had falsified the documents, and I was allowed to stay in France. The day that I was released from Fresnes prison, Claudia came to pick me up in a Volkswagen Beetle that Alain Guillerme was driving, with Danielle there as well.

The same day, I hugged Félix again and together we wrote the text of an appeal against repression in Italy, and against the historical compromise between the Communists and the Christian Democrats. The appeal obtained the support of Michel Foucault, Gilles Deleuze, Roland Barthes, Julia Kristeva, Philippe Sollers, Maria Antonietta Macciocchi and Jean-Paul Sartre, among many others.

In Italy, this produced a very strong impact, and the Italian intelligentsia reacted by expressing quite contrasting positions. Intellectual dissent was manifested for the first time as an international phenomenon able to create opposition with the same strength to western capitalism and to Soviet oppression and real socialism.

The appeal opened the way to the meeting against repression that took place in Bologna in September that year. That meeting was an extremely important event. Tens of thousands of people came (someone said a hundred thousand, but I could not calculate it). Enormous assemblies took place, meetings and street performances, improvised speeches and concerts. It was an explosion of joy and rage, but it also signalled in a certain sense the end of the history of the movements in Italy, opening the phase of the terrorist drift and of the State intervention aimed at wiping out dissident social forces.[4]

People came to Bologna in those days as if expecting a magic word, capable of opening the way to a new history, an egalitarian and libertarian history that would be at the forefront of the times that were about to come. It was as if everyone was there to hear the sounds of the era that was arriving, and to find the magic formula able to avoid the backlash, violence, catastrophe, isolation and defeat of any solidarity.

We did not succeed in finding the magic word.

We certainly made some mistakes. Perhaps we also made a mistake with the July 1977 appeal. We had placed State violence and repression at the centre, and had insisted on the right to dissent, while we probably should have insisted much more on the affirmative, creative character of the movement.

This way, we still couldn't have changed the course of history that had been prepared by a furious capitalist counteroffensive on an international scale, and the Thatcheresque counter-revolution on a global scale, and the attack on all forms of working-class life. We would not have changed history, but perhaps we would have promoted the metamorphosis of the rebels into autonomist innovators. In the following years, I saw Félix mostly to discuss what there was to do in order to help the political expatriates who were coming from Germany and Italy. Throughout the 1980s, the winter years, Guattari's own principal public engagement was aimed at denouncing political repression, and defending what had been won during past struggles.[5] But the philosophical creativity of Félix Guattari, in the books he wrote with Gilles Deleuze and in those he wrote alone, did not suffer at all from the consequences of the situation in which we had come to find ourselves, where we somehow had to defend our past, and our own possibility of survival.

Félix Guattari's philosophical creativity succeeds in delineating a rather broad panorama of what our strength could today encompass. In this sense, he sang the song of times that had to come.

****

Félix died in 1992.

The collapse of the Soviet bloc, the diffusion of ethnic-religious conflicts, the devastating deployment of the monetarist wave defined the horizon of the 1990s. After his death, I followed the evolution of the final decade of the century, considering rhizomatic thought as a map and trying to see the trace of the real in continuity with the lines on the map.

In continuity, not in analogy, because rhizomatic thought is not a calque, but a rhythm, a mode of functioning, a style. A rhythmic map, if I may say so.

With this book, I would like to reconstruct the rhythmic map of Félix-thought, and cause harmony to resonate among the chords, the refrains and the dissonances in the contemporary planetary rhapsody starting from that map.

Part I

2
The Happy Depression[1]

The 'Félix machine' is situated at the point of maximum expansion in the blossoming of a nomadic, provisional community. It also accompanies its dissolution.

It doesn't work in a linear fashion, and therefore it cannot be considered a volition-driven, mono-planar machine. If we want to see it in action, we also have to remember the people, the encounters, the conferences, and then we have to remember the images; we even have to repeat Félix's gestures, the movement of his hands and eyes. We have to find a hypertextual, translinguistic mechanism that would resemble his way of treating linguistic and emotional matters. We should think back to the mornings and evenings of times past, to the meetings that we hadn't found the time to attend. And then we have to understand how all these things could have disappeared.

*****

We have never elaborated philosophically the experience of depression. In fact, we have foreclosed it and made it shameful, as if it were something that cannot be addressed in public.

What a happy, *felix* hypocrisy.

In Félix's work, depression appears under the rubric *the winter years*.[2] But we cannot reduce it to the winter years, for it's not the winter's fault. Desire is cruel, and so are autonomy, beauty and the irresponsibility of dancing.

Depression presents us with the bill.

The Subject can't refuse to pay the bill, nor can the singularity.

Depression *is* the bill.

*****

What is depression?

Depression has something to do with the problem of sense. Sense is the production of desire. Desire is dissipation.

I am not talking about the subject here, but of the singularity as a living, conscious and sexual organism, as an organism whose existence is condemned to dissipation.

Depression is the *fallout*[3] of the megalomania implicit in the construction of sense. The enunciation that confers sense to the world, and not simply limited to function in its economy or to answer already established questions – this enunciation is a struggle to raise new questions that happens on a cliff overlooking the abyss of depression.

The lightness of existence is a scandal: the word is born of words and suspends all relations, the organism, society, sexuality, love, the others. Even time is suspended. But how long can such a dance last?

This is a difficult dance, in which you may fall.

*****

We will not again start the process of creating sense without facing depression, without paying our bill.

I want to talk to my friends, to Félix's friends. Every time we had a chance to be together we would wear our holiday clothing because being together was always a holiday.

But we also know that there was a whole other story that we kept hiding away.

This is the story of depression, and Félix chose not to tell it.

This depression, we called it Joséphine.

Joséphine was Félix's partner from 1986. Félix came to see me with her in via Marsili, in the spring of that same year. Joséphine was fragile and unstable, sexy in her spiky heels and black-and-white polka-dot skirt.

*****

One day, I went to see Félix. He looked more mysterious than usual, he smiled sadly and told me, pointing at the medicine bottles on his desk: you see, I prescribed all these medicines for myself just to become stupid.

Then we went out to see the Tinguely show at the Beaubourg.[4]

He sat down on the floor looking at the cogwheels and the whole cosmic-archaic-mechanic machinery, and kept staring at that emptiness for many minutes.

Felix did not pay attention to depression, neither as a philosopher, nor as a psychoanalyst.

And we can easily understand why. The methodology [*démarche*] of the *Anti-Oedipus* is not easy to reconcile with the possibility of delving into depression. Depression is not just a condition among others, in which a machinic unconscious is assembled, made of existential and chaosmotic fragments proceeding from anywhere to everywhere else.

The *Anti-Oedipus* does not know depression; it continuously overcomes, leaping with psychedelic energy over any slowing down and any darkness.

*****

But there is a time for depression.

And we shouldn't underestimate its cognitive potential. Of course, depressive intelligence is paralyzing. It gives rise to the feeling of an immense discrepancy between sense and the everyday, and to the feeling of the dispersion of desiring energy in the contact with the world in time.

The investment of desiring energy is the origin of sense.

The world in time is the dispersion of this energy.

The activity of the philosopher is the fabrication of concepts.

Concepts are a concretion of desiring energy.

They give sense to the world, singularizing it and projecting desire in such a way that it might become the world.

But the world flows and disperses itself.

Concepts no longer grasp configurations, they no longer project sense in a shared reality.

The dissolution of shared experience is the entropy we have lived starting from a certain moment in time – we could say from the very beginning, but more intensely, painfully, rapidly, catastrophically during the 1980s. *The Winter Years*.[5]

The winter years are this, they are the time when the contradiction dissolves.

The places of existential trajectory become poorer, more rarefied and artificial, because concepts are losing their hold. Or maybe concepts lose their hold because the sites of our existence have become more rarefied, poorer, more artificial and ungenerous.

The desiring community is a chaoid, a provisional organizer of chaos, a fragile architecture composed of shared happiness and common vision.

The dissolution of a desiring community inaugurates the painful spreading of chaos, of an absence of sense. It is 'kipple', Philip Dick would say, cosmic garbage, what remains of the world when lovers are lost and friendship dies out.[6]

*****

Let me be absolutely clear: the desiring community has nothing to do with a community of origins, with a people, an ethnicity or a nation.

A community of origins is grounded in a 'belonging': differences are nothing, all that counts is identity. Quite to the contrary: in the desiring community there is no belonging, but only the encounter of singularities. This encounter cannot be structured in a stable way; it is constantly reborn out of the pleasure derived from the common enterprise of conceptual projection.

And we have known very well the desiring community. We have known it in the sunny mornings, when a friendly body came to wake us up stroking our body under unexpected sheets.

We have known it in the street barricades, among the burned cars, trying to prevent the advance of the military vehicles.

We have known it, screaming with joy, in the sudden encounters in train stations submerged by the fog.

We have known it in the nocturnal sound of a flute accompanying a small crowd stopping while waiting for the last subway ride.

Anti-Oedipus is the manifesto for the provisional and nomadic desiring community.

With *Anti-Oedipus*, modernity's conceptual machine arrives at the point of catastrophe.

The desiring community and philosophical activity camp out under the same tent.

In the introduction to *What is Philosophy?* the authors say:

> What does *friend* mean when it becomes a conceptual persona, or a condition for the exercise of thought? Or rather, are we not talking of the lover? Does not the friend reintroduce into thought a vital relationship with the Other that was supposed to have been excluded from pure thought? (*What is Philosophy?*, 3–4)

At a certain point in our lives and in the history of the *Anti-Oedipus* generation, which is the generation of 1968 – and there is no better way to define it – a movement of dispersion was unleashed.

The words available to us were no longer able to seduce the world's ears, the concepts didn't grasp reality any more, the movements of men could no longer understand the essence of technology and language.

*****

I remember that in the 1980s Félix often scolded me because I was no longer involved in some kind of political militancy. I had stopped considering myself a political militant already in 1977. For me, the movements of 1977 had actually been a critique of militancy and a gradual overcoming of modernity's politico-existential conceptuality. The movements of 1977 always appeared to me modernity's final spasm, the beginning of a transition – and then I didn't know how long and painful it would be – toward a new world.

For me, militant will and ideological action had become impotent.

This is why Félix reproached me, jokingly but not really. He wanted me to get involved with the Italian Green movements, as he himself – generously but somewhat ineffectively – was doing in France. He wanted me to run for office, in Italy. I said no, even if I went with him to several meetings he had with the Parisian ecologists and other militant groups. I found those meetings quite senseless. Félix had nothing to do with militant meetings anymore, with political action. There wasn't any other possibility, for political action, than pure resistance. But resistance is hopeless, because when you resist you are actually defending conceptual configurations that have already lost their grip on the world. When you resist, you replace desire with duty, and this cannot work if we believe in a kind of creationist process.

Resistance is the opposite of creationism.

Félix knew this, I am sure, but he never said this much, not even to himself, and this is why he went to all these meetings with people who didn't appeal to him, talking about things that distracted him and making lists of deadlines and appointments.

And then he would run off, adjusting his glasses to consult his overflowing daily planner.

And here again is the root of depression, in this impotence of political will that we haven't had the courage to admit.

I am not saying that depression has a political origin, nor do I want to forget Joséphine. I just want to say that depression is born out of the dispersion of the community's immediacy. Autonomous and desiring politics was a proliferating community. When the proliferating power is lost, the social becomes the place of depression.

The community I am describing is not based on the origin, on the past, on the ties with the earth, the blood or identity. It is the provisional community that founds itself on the possible, on desire, on intention, on the promise, on expectation.

Provisional communities, this is what I am talking about. The movements of 1968 started a provisional, nomadic community which then slowly decomposed and dispersed itself.

The dissolution of sense is parallel to the dissolution of an immediate recognition.

Desire and depression are united by imperceptible ties

It seems to me that in the notion of desire there is a hidden relation with the Christian and Romantic traditions – a relation that is not completely rescinded in the thought of Deleuze and Guattari. This is an idea which is already present in the discourse of Georges Bataille in the notion of *expenditure* (*dépense*): a transgressive excitement, a mobilizing tension, an energetic investment in the future. Desire is the utopian tension projecting consciousness onto the world; it is the origin and motivation for the projection into the world. Depression is rooted in this tension.

By that I don't mean that we should abandon or liquidate the notion of desire that in *Anti-Oedipus* was the productive power of the unconscious. Still, we can't ignore the fact that desire means tension, and that tension is destined to fall, to die down, given the irreducibility of existence, the decomposition of organic matter, our being-for-death.

The concept of desire is linked to a youthful utopia deriving from Romanticism, a utopia which we shouldn't disavow but cannot worship either.

Anti-Oedipus talks about the productivity of the unconscious. The world emerges from the desiring factory as a flow of shared illusions. This reality is the product of a shared desire.

But Hindu thought calls this shared illusion *maia*, a notion that should be integrated, I think, with the concept of desire. Buddhism, which stems from the Hindu tradition and develops it further, elaborates the notion of enlightenment or awakening, which is similar to *nirvana*: the clear understanding of impermanence, of the empty and delusional character of reality. (In the Western tradition we also find some references to this modulation of energetic investment: Stoic thought looked for happiness in *ataraxia*, in the suspension of desire, in the active interruption of worldly tension).

Enlightenment is the understanding of the nullity of *samsara*, of experience, and this is why the enlightened can live desire without getting attached to it: he rejoices in the illusion, but his soul does not depend on it.

'Even the most frightening monsters are only the hallucinations of your mind': these are the words of the *Bardo Todol*.[7] Desire is the force giving life to demons and visions. It is a shared hallucination.

In their last shared work, *What is Philosophy?*, Gilles and Félix began the elaboration of a senile utopia that perhaps we should consider more deeply.

Today we need a senile utopia that would be able to integrate the youthful utopia of desire.

The senile utopia of friendship is a sharing of experience but also an understanding of its hallucinatory and thus impermanent character; it is the clear vision of its own nullity, the dissolution of the dependency and the attachment translating desire into depression.

This is the way to *nirvana*, the contemplation of nothingness which in Taoist tradition is called *wu wei*.

This is the condition for being able to live desire while transcending it at the same time. This process of transcendence is not purely intellectual, but also experiential, aesthetic and sensual, and is founded on the understanding of the impermanence of both body and soul.

3
Integrated World Capitalism
(From the simulation of global war to the reality of globalization)[1]

The early 1980s

While President Reagan, the incarnation of the imaginary of Hollywood, initiated among glitz and glamour the frenetic simulation of Star Wars, on the other side the cadaverous Brezhnev regime thundered against Afghanistan and grimly threatened the workers' revolt of Gdansk with tanks,[2] but without having the courage to confront it directly.

A few years before, in one of his extraordinarily insightful essays, Zbignew Brzezinski had discussed the power of simulations in international politics.

The internal system of armed dissuasion between the two nuclear powers always was a giant game of simulation. At the start of the 1980s, this game reached its high point. The installation of Soviet SS20s, and the parallel installation of American Pershing missiles along the European border that divided the two blocs, produced a formidable ideological effect (or maybe it would be better to say a psychological effect, or psycho-chemical, or psycho-social). The imaginary energies, which had in the preceding decades been directed toward the progressive transformation of the political equilibrium toward an opening of new social perspectives, were all completely and suddenly shut down and fell back onto war. Military expenses increased in an unprecedented manner. After all, the Hollywood President directly represented the interests of economic groups that were invested in the military industry, and the living cadaver of Brezhnev based his power on the social force and productivity of the war apparatus.

One geopolitical bloc played its simulation against the other geopolitical bloc. But behind the geopolitical game another game was being played, which was the real game.

Many years later, when the Soviet giant had already collapsed under the economic, military and ideological pressure from the Western bloc, someone in the Pentagon revealed that its Star Wars project, to which the Reagan administration had devoted enormous financial resources and decisive strategic value, had only been a bluff. The missiles could not have effectively protected American cities from a hypothetical Soviet attack. But the bluff worked.

For several years, from the invasion of Afghanistan to Gorbachev's rise to the presidency of the USSR, it seemed as if the conditions for a direct confrontation between the two blocs were truly in preparation. In reality, both blocs were jointly playing their super-simulations before humanity: by moving economic and creative energies toward war, they blocked the process of social transformation. The Soviet bloc tried a final defensive manoeuvre, freezing itself in an attempt to freeze the outpouring of the different sorts of separatist, nationalist and religious tendencies that in 1989 finally aligned themselves with the democratic movement if only to eviscerate it later.

Meanwhile the Western bloc initiated a process of redistribution of social wealth in an anti-proletarian direction and a productive restructuration based on new technologies.

Behind the screen of the simulation of a totally asymptotic war (a war always promised and never unleashed that still produced its effect on society and in the imaginary), what was really produced were the effects of a frightening (perhaps irreversible) destruction in the planetary fabric of society. Military superproduction, even if it did not lead to a direct confrontation between the two blocs, increased the occurrence of wars and the militarization of entire areas of the planet.

The war that Saddam Hussein unleashed in the 1980s against the Khomeini regime was financed and armed by Western countries. The USSR itself waged war in Afghanistan. Everywhere in the world during these years the ground was laid for a militarization that destroyed the economic and environmental resources causing a feeling of terror and aggression that poisoned inter-ethnic and international relations. Only in the 1990s have we seen the slow dispersion of the effects produced by this injection of aggressive adrenaline into the planetary body and mind.

The vast majority of European intellectuals during these years concentrated their attention on the possibility of total war and of the seemingly imminent danger of a confrontation between the USA and the USSR.

In sharp contrast, Guattari focused his attention on the planetary integration of capitalism:

> The great crisis of this end of millennium appears to be announced as the conjunction of all possible convulsions. There would be no end in enumerating the fields that are adrift or that are evolving in catastrophic ways (energy, employment, ecology, demography, international relations ...). Without any doubt, there is no reason to expect a nuclear holocaust, but perhaps we are now already engaged in a kind of new Hundred Years War ... (*Piano sul pianeta*, 91)

These are the opening words in the conclusion of a little book that was first produced through a militant network in Paris, and was published in Italy in 1982 with the title *Piano sul pianeta. Capitale mondiale integrato e globalizzazione*, and republished by Ombre corte in 1997.[3]

The concept of integrated world capitalism at that time had a quasi-scandalous ring to it. At the very moment that the fracture, the opposition, and the threat of war seemed the most acute, Guattari focused his attention on a process that today is easy to recognize and understand, but that then seemed counter-intuitive: the process of globalization, of the integration of markets and circuits of production. Many current discussions about globalization were anticipated by this small book, starting with the prediction of a crisis in the social-authoritarian field.

****

A truly progressive workers' perspective never deemed real socialism to be an alternative to capitalism. The entire experience of the social-authoritarian regime was based on substantial acceptance of the hierarchical rules of capitalist economic theory, and the social basis of this regime was the conservative resistance of feudal, military and ideological bureaucracies to the expansive and progressive dynamic of global capital.

Both in the Soviet and in the Chinese empires the retrograde defence of a system of privileges has been valorized for decades, rather than the experimentation with forms of collective property. After the decline and collapse of these tyrannical regimes, it is becoming evident today that their function was essentially conservative and bureaucratic in nature.

The end of this authoritarian model freed up enormous productive and market potentialities, ushering in an extended phase of instability, violent transformation, and the fear of aggression and violence. At the same time, we know well today that the end of social-authoritarian regimes did not mean in any way the start of the reign of democracy, as the Reagan apologists had proclaimed. Democracy did not enter in any way into the end of social-authoritarian regimes which, in the 1990s, acquired characteristics of Nazi-Communistic regimes.

After 1989, with an exception made for marginal cases like Poland and the Czech Republic, in the Soviet-bloc countries and even more patently in China, a totalitarian regime was put in place that was even more ferocious, more corrupt, more arbitrary and more bloodthirsty than the ones that previously had proudly brandished the hammer-and-sickle on a red background.

The empire of Evil is succeeded by the empire of even Worse.

But for the West, none of this matters. The only thing that really matters is the collapse of the feudal-bureaucratic barrier, the military weakening of the feudal empires of the Orient and thus the opening of immense markets, and the creation of new, gigantic divisions of production in the international system: divisions in which manpower has minimal costs and social peace is guaranteed by the criminal violence of the Nazi-Communists in power.

****

Precisely when Western intellectuals and an infantilized European public opinion brought about a form of nervous exhaustion through fear of nuclear holocaust, Guattari went straight to the point and spoke about integrated world capitalism.[4]

Twenty years after this little book's publication, Guattarian intuition proved itself to be the order of the day. We must return to this reasoning and develop it.

What differences are there between world-wide markets and globalization? The two terms do not define the same thing, or the same process.

The process of creating a world-wide market is defined by an increasing exchange of goods through different zones of the planet, by an increasing integration of markets, and consequently by lifestyles that are linked to consumption. The increasing rates of national outputs of producer nations are consumed in geographical areas quite distant from those in which they are produced.

The process of globalization includes an integration of cycles of production. Increasing rates of production are the result of a planetary assemblage, of a horizontal integration across different moments of the labour process (planning, partial manufacturing of kinds of merchandise, assemblage and testing, styling, advertising) that take place in different parts of the planet.

While a world-wide market implies the mobility of finished goods, the process of globalization instead involves a true and precise deterritorialization of the process of production. In the phase we are defining as globalization there are no longer any relations between financial investment and control over production. Whoever invests capital is interested in knowing how the company's stock in which one invests will fare, but one is not expected to know what goods the company produces. The divorce between exchange value and use value is definitive. The circulation of value diverges completely from the material circulation of goods produced. The product that is especially exchanged, above all, is information.

It is evident that this transformation that we are calling globalization is made possible by the diffusion of technologies of communication and virtualization: the process of production is in large part dematerialized, that is, modes of information are produced. Generally the productive sectors, in which materials that mobilize physical energies of the classic industrial kind must be handled, are situated in more peripheral locations in the international economic system, where the workforce can be had at minimal cost.

****

The whole process that today is being extensively deployed was conceptualized by Félix Guattari precisely in this essay.

After having discarded decisively the fear of the prospect of nuclear war, and having recognized its character of pure simulation, after having implicitly foreseen the desegregation of the social-authoritarian systems and their economic assimilation to integrated world capitalism, Guattari furnished some elements for analysing this model.

The premise of this analysis is contained on the first page of *Piano sul pianeta:* 'Capital is not an abstract category, but rather a semiotic operator' (3).

What does this mean? It means that while the labour process is fragmented, extended, recomposed and decomposed through deterritorializations of all kinds, the process of valorization integrates all the fragments of capitalist production not only (not simply) through the abstract functioning of the laws of value, but also through the concrete, direct action of the technologies, allowing the instantaneous movement of information.

> Abstract Marxist values overcoded the aggregate of human labor concretely dedicated to the production of exchange values. But the actual movement of capitalism tends toward the point at which use values become exchange values, and all productive labor depends on mechanization. The poles of exchange themselves are now on the side of mechanization since computers talk among themselves from one continent to another and dictate to the managers the terms of the exchange. Automatized and informatized production no longer gains its consistency from a basic human factor, but from an element of mechanized continuity that traverses, surrounds, disperses, miniaturizes and recuperates all human functions and activities. (*Piano sul pianeta*, 10)

When Guattari says that capital is a semiotic operator, he means therefore that the pervasiveness of the capitalist model no longer depends solely on an effect of abstract overcoding that manifests itself especially in the moment of exchange, but also depends on the technologically mediated integration of different moments of manufacturing: planifying moments, techno-scientific moments, informational moments, material moments, and so forth.

Guattari did not allow himself to be distracted in the least by the game of simulation that was taking place on the world stage at the start of the 1980s. Rather, he pointed directly to the long-term trends, and prefigured in this way the process that we see developing in the 1990s:

a) the imposition (or rather the infiltration, penetration, invasive proliferation) of the capitalistic model understood as semiotic operator, as rule of generalized transcodification;
b) the proliferation of the margins, both in the form of residuals, re-emergence, reterritorializations (the identitarian obsession, nationalisms, forms of integralism, tribalisms), and in the form of minorities, autonomies, deterritorializations (subcultures, provisional communities, cultural contaminations).

Therefore this document on integrated world capitalism impresses us today as a visionary text and helps us understand something about what is happening in the planetary economy and psycho-chemistry.

4
Planetary Psychopathology

Félix Guattari died one night in August 1992. This was when the illusion of peace evaporated which, in 1989, had accompanied the collapse of the social-authoritarian bloc and the end of the Cold War. The Gulf War in 1991, and then the development of the Yugoslavian situation all the way to the explosion of the Serbo-Croatian war, opened a new world stage. No longer was there a bi-polar front, and therefore military conflicts were no longer controllable in a centralized manner. Conflict happened along lines irreducible to global order or a unitary strategy. Lines of ethnic, religious, Nazi-nationalistic and tribalist sorts overlapped with the lines of the planetary economic conflict. On the economic level, the interests of integrated world capitalism (which was pushing toward globalization) ferociously compressed the living standard of the masses in the world's impoverished countries, over which the international organisms controlling the economy imposed a free-trade choice that functioned in a catastrophic way, enriching a globalized bourgeoisie (or a virtual class integrated into global capital). The effects were evident in the final years of the 1990s with the collapse of Russian society, and the economic crises of countries in the Far East, from Thailand to Korea and even Japan.

In his final book, *Chaosmosis*, written shortly before his death, Guattari wrote:

> Generally, one can say that contemporary history is increasingly dominated by rising demands for subjective singularity – quarrels over language, autonomist demands, issues of nationalism and of the nation which, in total ambiguity, express on the one hand an aspiration for national liberation, but also manifest themselves in what I would call conservative reterritorializations of subjectivity.

> A certain universal representation of subjectivity, incarnated by capitalist colonialism in both East and West, has gone bankrupt – although it is not yet possible to fully measure the scale of such a failure. (*Chaosmosis*, 3, translation modified using the Italian version)

The disturbance into which the planet entered at the start of the 1990s after the brief hope for peace following 1989 was of a new sort, having little or nothing in common with past economic crises. First of all because this exploded in conjunction with the rapid and apparently uncontainable spread of planetary civil war, of a tribal war waged with ultramodern weapons by everyone against everybody else.

The traditional categories of economic analysis say nothing about the factors that appear to be decisive in today's world: the social investments of desire, their psychopathology, the intersection between economic and semiotic flows, the disturbances occurring in certain fundamental cognitive functions because of the mutation of the natural and infospheric environment.[1]

In order to understand what happened in the politics and economy of the 1990s, it is indispensable to make use of the conceptual instruments of psychopathology and of schizoanalysis, since the crisis is above all an incompatibility between the fantasmatic projections of different social, cultural and national formations.

What were the innovative traits of this extraordinary abyss? The 1980s and 1990s were a period of great economic innovation and of an enormous increase in global productivity and wealth. But these were simultaneously the years of the systematic destruction of the natural planetary environment and of the psychic environment in which humanity lives and communicates.

Admitting that the concept of neo-liberalism might be adequate for understanding the complex dynamic that subjects the internal mental and material system of production to the dictatorship of immediate private profit, one can then say that in the transition period between industrial and post-industrial phases, neo-liberalism legitimated a process of destruction not only of actual but also of future resources. This process occurred by frenetically stimulating debt, overconsumption and competition, and by forcing the collective psyche to undergo a competitive stress, the effects of which produce depression, panic and aggression.

In the course of the 1980s, states, communities, corporations and individuals, in order to participate in the game of competition, were pushed to place suicidal bets on the gaming table, risking their future existence and resources.

Thus, what appeared on the horizon at the end of the 1990s is a form of planetary civil war that fragmented the globe into a series of innumerable local conflicts, all incapable of putting forward any universal values.

The cultural devastation produced by neo-liberalism has upset social investments of desire, provoking a drought in productive social creativity and determining a true emotional plague, the aggression of everyone against everyone else, an obsessive fear of contact, a wave of Nazism without ideology, a purely visceral racism. The fragmentary war that has begun does not seem reducible to a political plan, and thus risks not dissipating before having produced all the destruction of which it might be capable.

The majority of humanity – the eighty per cent of the planet that must be satisfied with twenty per cent of the resources – finds no other prospects than that of a mortal conflict in which the poor grab bread and territory away from other poor, while the centre of real power seems to have become inaccessible, untouchable and even unrecognizable.

But even in the rich metropolises of the world, civil war is evident, in forms of mafia-like criminality, of armed gangs that fight over territory and rackets, and of creeping or outright racism.

The destruction of the environment and economic psychopathology

At the threshold of the 1980s, punk had proclaimed the abolition of the future.

Humanity no longer had any reason for remaining together. This was the feeling, the premonition. Nevertheless, the politicians, journalists, intellectuals of the time all hastened to declare that, with proletarian and autonomist extremism finally defeated, with the workers' political strength finally eliminated, with utopia finally stamped out, an era of unlimited enrichment could begin.

Officially the 1980s were the years of development and freedom.

It matters little that competition was exalted as the noblest feeling, and so violence of one against all others was thereby sanctified; it matters little that profit became the ruling point in culture and in communication. So, the arrogance of the ignorant became widespread and promoted as a way of life and of achieving success. This decade was an age of capitalist insurrection, and thus it officially became the age of civilization, freedom and progress.

The politics of the Western neo-liberal elites was founded on the creation of a new bourgeois non-productivity, on an accelerated destruction of resources aimed at financing the illusion of well-being, on the accumulation of public and private debt, on the displacement of economic energy from the production of goods to financial speculation, and on an absurd militarization of production in order to bring the economies of Eastern-bloc totalitarian countries to their knees and to reintroduce artificially industrial production.

At the end of the decade, the results were evident: we witnessed the collapse of Soviet social-imperialism, but also the collapse of all political and economic balance in the entire Euro-Asiatic territory, from Berlin to Vladivostok. The generalized return of nationalisms, tribalisms and aggression followed the rapid collapse of the illusions produced by neo-liberalism and the destruction of any form of identity.

The mixture of generalized competition and planetary militarization has brought about an ungovernable multiplicity of archaic, tribal, mafia-like and religious conflicts fought with ultramodern weapons. To judge the 1980s, one has to situate oneself in the perspective of the consequences that this period produced in the living body of the planetary society, of the human mind, and of the possibility of survival on the earth.

****

Rio de Janeiro, June 1992, the Conference of United Nations on development and environment, also called the Earth Summit.[2] Confronted with the alternative of ecology or economy, the Western political leaders, especially the American ones, stated the truth: between the danger of a slowdown in economic growth and the destruction of the planetary environment, we choose without any doubt the second perspective.

The alternative between returning to a livable dimension of the environment and maintaining the rhythm of development and consumption to which Western public opinion has become accustomed is a chokehold that the political class is absolutely unable to loosen.

> Third world debt and the destruction of the natural environment are closely intertwined phenomena, even if the nature of their relations has changed in the course of the last decades. During the 1970s, the developing countries went heavily into debt in order to

> finance huge projects that were revealed to be veritable ecological nightmares. ...
>
> Ever since the explosion of the debt crisis in 1982, Latin America has sent each month (and over one-hundred and eight months) four billion dollars to the North. Even sub-Saharan Africa, stuck in the mud of its poverty, one way or another paid a million dollars a month to reimburse its loans ... While countries are falling into ruin to continue to export, the World Bank and the IMF urge them to impose severe cuts in public expenditures. Budgets for protecting the environment are inevitably the first to be sacrificed ...
>
> The gravest ecological consequence of the debt is perhaps the massive deforestation that it has encouraged. Trees are cut down in the most irresponsible way to build furniture, window-frames and chopsticks. Sometimes forests are simply levelled and turned into pasture land for cattle whose meat is sent to furnish fast-food restaurants in the North. The levels of indebtedness are directly connected both to the extension and to the rhythm of the destruction of the forests; the more the debt rises, the better are the conditions for the dump trucks and bulldozers. (Susan George, 'La dette se paie en nature', ii, vi, citation revised to conform to original)[3]

Throughout the 1980s, the ruling groups promised possibilities of unlimited enrichment for the mass lumpen bourgeoisie that has formed in the northern sector of the planet as well as in some countries of the Third World. The media and advertising bombardment of the 1980s has served to inject into the collective brain the stimuli caused by hypertrophic needs. It seems rather difficult that this small and decisive part of humanity, corresponding more or less to Western Europe, the United States and Japan, will ever accept a change that would endanger the motivation frenetically inculcated over a decade as the only one that might provide meaning and identity: economic motivation and competition.

In 1992, when Bush (Senior) appeared in Rio de Janeiro to respond in an arrogant way to the accusations raised against him by several participants, his speech had the strength of common sense, of the self-evident. Many Third World countries, the organizing forces behind the summit, and several European countries reproached the American administration for not having accepted in any manner to start planning a reduction in toxic emissions in the course of the next decade. But what does the reduction of toxic emissions mean? It means a slowdown in the development of the automobile industry, the main factor in the atmospheric pollution in the major metropolises and the entire planet. After ten years of reduction of the workforce in the automobile industry, faced with the severe recession that struck Western markets, the capitalist economy could not even consider the hypothesis of a reduction in automobile production. The same configuration of the Western imaginary made it impossible even to think about an organization of civic life capable of avoiding the devastation of our planet's environmental resources.

****

Félix Guattari taught me to see social processes as the production of unconscious processes, and to see the unconscious as the laboratory in which the stages for social action are produced. There is no need to think of power as a cold machine of decision and will.

When one uses words like euphoria or panic or depression to describe the behaviour of the stock exchange or the markets, we must not think that this only is a question of metaphors. It is also a question of an adequate description of the psychopathology that traverses the social mind in a situation of informational overload and competitive stress.

> Sometimes, for example, economic semiotization becomes dependent on collective psychological factors – look at the sensitivity of the stock exchange to fluctuations of opinion. (*Chaosmosis*, 1)

The increasing mental nature of production has exposed the economic system more and more to the psychic storms that are crisscrossing the collective mind. And on the other hand, the economy has caused competitive sentiments that have been transformed into anxiety, and at times degenerate into forms of panic. Financial and psychic flows are closely interdependent. This is why we must study the political, economic and social catastrophe beginning with an analysis of the social investments of desire.

The economic finalization of desire, the harnessing of psychic energies, the competitive acceleration of all relations become comprehensible within the same temporal context as the media epidemic of AIDS. Since true catastrophes never occur alone, the ideological dictatorship of neo-liberalism in the 1980s was accompanied by the spread of AIDS. The two phenomena are quite different, of course, but perfectly integrated on the level of the social imaginary and in their effects on behaviour.

In a fanatically economic and competitive society the body must be at work around the clock. The constant mobilization of energies produces a kind of paralysis of the erotic body. Eroticism is represented either as gymnastics or as perversion, but in both cases the perception of the other's body is increasingly sterile. In the course of the 1980s, AIDS produced a convergent de-eroticizing effect.

The other's body tends to be redefined on the level of psychic perception as a danger and as a potential factor of contagion. Social relations then cool down, are de-eroticized, and are increasingly transformed into functional relations.

Are we not determining the conditions of a psychic catastrophe?

Are we perhaps cancelling pleasure on this earth?

5
Postmediatic Affect

The Italian radio movement, and particularly the experiment of Radio Alice, which expressed a Dadaist, schizo impulse, had an influence on the start of the Parisian free radios, and Félix was very active in that history. In France, the free radios had an even greater disruptive meaning, if that was at all possible, than in Italy. The traditional State centralization in France reacted with great suspicion to the blossoming of a multiplicity of voices and, for the first time, minority and underground cultures were able to have a voice. Radio Soleil broadcast music from the Maghreb, and Radio Tomate reported the ecological battles and the occupation of houses by young proletarians in the Parisian suburbs.

In 1977, there appeared the French edition of the small book *Alice is the Devil, Free Radio Alice* (*Alice è il diavolo – Radio Alice Radio Libre*), and Félix wrote its preface, entitled 'Des millions et millions d'Alice en puissance' ('Millions and Millions of Potential Alices'). In the radio experiences, Félix saw the proliferation process of enunciative agencies destined to cause the explosion of the mass-media model. When Félix died at the beginning of the 1990s, the fever of the World Wide Web had just begun that would, in the following years, transform the global communication system, introducing a rhizomatic notion that eroded the centralistic principle of the twentieth-century mediatic system. Conceptually, the free radio movement had anticipated the post-mediatic tendency that the Internet is today bringing to maturity.

****

Guattari had always played fearlessly with technology, yet he personally did not use a typewriter and wrote with a cheap ball-point pen that can be found in any Parisian cigarette shop. Machines, gadgets, assemblages and bodies without organs were at the centre of his thought; they were the code informing his language, the surrealist décor of his imaginary.

European critical thought has always had problems in its relations with technology. Benjamin was able to introduce a philosophical perspective regarding the techno-communication machine, and he could recognize that the industrialization of culture modified the very conditions of aesthetics. But the main current of critical thought has always considered communication technologies to be an instrument of social culture's subjugation to the dominant ideology or as messages inducing consumerist and conformist forms of behaviour.

Not Guattari. His lookout post had always been the net, even when the word did not have the meaning that it has for us today, even before the World Wide Web existed and the Internet started making newspaper headlines.

In a small pamphlet entitled *The Holy Fools*, Richard Barbrook insults Deleuze and Guattari's thought in the name of labourist and statist orthodoxy, arguing that rhizomatic thought is akin to Californian ideology, that is, to the kind of neo-liberalist thought that sings the praises of high-tech capitalism.

Rhizomatic or techno-nomadic thought does, in fact, have one thing in common with the apologies of high-tech capitalism: it takes a perspective of becoming and not one of conservation, and thus it succeeds in understanding the logic of mental labour typical of network technology and of pan-capitalism, in seeing things from the inside and not from the viewpoint of the State, national sovereignty and past identities.

Deleuze and Guattari's techno-nomadic thought analyses contemporary capitalism as a semiotic flow and situates the task of critical thought on this plane. Thus if today we wish to reason in terms of the molecular self-organization of neo-labour against capital-flow, we need to refer to schizoanalytical – not Marxist-Leninist – concepts.

Techno-nomadic thought is the determined and specific subversion of high-tech ideology, a kind of thought that is able to spread out along the same lines and same rhythms of semio-capital.

Félix Guattari wrote:

> The democratic chaos [...] conceals a multitude of vectors of re-singularization, attractors of social creativity in search of actual-

> ization. No question here of aleatory neo-liberalism with its fanaticism for the market economy, for a univocal market, for a market of redundancies of capitalist power. (*Chaosmosis*, 117)

For Guattari, the free radios were a general rehearsal for the emergence of these resingularizing vectors, of these attractors of social creativity. And in fact, in reflecting on this phenomenon twenty years later, we see very clearly that the free radios were an anticipation of the Internet model now representing the incarnation of what Félix called post-mediatic civilization. The experience of the free radios (and particularly of Radio Alice, which from start to finish expressed the awareness of specific techno-mediatic linkages represented by radio in its continual interaction with listeners) anticipated a process of techno-communicative self-organization prefiguring the end of the mediatic era. This awareness made Guattari a precursor of libertarian cyberculture.

Pressed by the diffusion of electronic communication technologies and, in particular, the Minitel experience that became prevalent in France during the early 1980s and represented the first example of a European telematic network, Félix managed to speak about the post-mediatic civilization starting to appear on the horizon. This would be a civilization in which communication flows are no longer directed from above toward a passive public and instead function as the densest framework for rhizomatic exchange through emitters situated on the same plane.

'The police got rid of Alice – its perpetrators were pursued, condemned and imprisoned, and its premises ransacked – but its work of revolutionary de-territorialization still goes on unabated, even affecting the nerves of the opposition,' Guattari wrote in the introduction to *Radio Alice Radio Libre* ('Des millions et millions ...', 241).

Radio Alice can be considered as the first experiment of deterritorialization of the telecommunication system, and of attack against the centralized media system.

Thanks to the free radios it was possible to understand for the first time a principle that the Internet has today propagated: the networked diffusion of communication is the privileged plane of social self-organization.

This seems to me the meaning of Guattari's words: Radio Alice was not an instrument of information, but a device for destructuring the media system, the trigger for a destructuring of the social nervous system that continued in the following decades, with liberating but also catastrophic and panic-inducing results.

*****

The becoming of communication is in a close relationship with the becoming of social perception and of aesthetics. In *Chaosmosis*, Guattari speaks of 'the new aesthetic paradigm' (*'le nouveau paradigme esthétique'*): the theme of his study is the mutation of the epidermis, of the contact zone between bodies, of sensibility.

What is aesthetics? In contrast to the prevalent understanding of Western philosophy, aesthetics is not only the science of an object's beauty. Aesthetics is also (and this is what is of greatest interest) the science of sensibility, of perception, the science of the contact between epidermises, and thus the science of the projection of worlds by subjectivities in becoming.

There is no social question any more essential than this one, because cognitive capitalism is an affection of sensibility more than anything else. In the sphere of globalized capitalism, the essential site of work, of economic exploitation, of psychic suffering and affective solicitation is the human mind and, more precisely, the body-mind relationship undergoing the pathogenic effects of informational overstimulation.

Capitalistic acceleration, the rarefaction of the contact between bodies, replaced by communication, planetary ethnic deterritorialization, the disagregation and collapse of traditional anthropological models: all of these act on the modalities of elaboration of the social mind and, above all, on sensibility. The emotional body is the site in which the most delicate and extreme of battles is being waged.

The mediatic bombardment of social attention produces brutal effects on sensibility. The imaginary is invaded by hyperfast monsters, mutagenic viruses of the collective psyche. And meanwhile, the unleashing of the mediatic epidemic taking the name of AIDS, a contagion that is essentially spread in the human psyche, goes hand in hand with the virtualization of relations among conscious organisms.[1]

This is the sphere of interest of schizoanalytic aesthetics: its problem is not beauty as an object of contemplation, but the way in which bodies perceive each other in the social field. In an era of displacement and migrations, of contaminations and integralisms, of nationalisms and aggression, an essential political problem is that of the semantics of social proximity, and thus of aesthetics.

In a certain sense, we can say that the semantics of social proximity are the privileged object of Guattarian aesthetics. And to understand how these social semantics are transformed, one must understand how art acts in communication, and how communication acts on the collective mind.

*****

Art is a semiotic action of a deterritorializing kind. Gestures and semiotic signs modify the relation between the sign and its context, its function, and the conditions of common interpretation. In traditional societies that are strongly territorialized, the artistic gesture is exceptional and sacralized. But in the twentieth century, when the inflated weathervane of generalized aestheticization became unhinged, something broke forever in the status of art. The loss of the aura is the disappearance of the exceptional and unique character of the work of art. The industrial reproduction of the artistic sign opened the door to semiotic inflation. Cinema, sound recording, television, advertising, digitalization and finally the automatic creation of artworks all dispersed the authorial aura.

With the dynamics of 'new economy' capitalism, the whole process reaches its term and its reversal. We were used to thinking that art deterritorialized and economy reterritorialized. Now we see that economy has subsumed art as a factor of perpetual deterritorialization and of valorization without territory. Chance, once a model of communication available only to the artist, has now become the predominant semiotic regime. Economic values are exchanged according to rules that change from day to day, while in the past this was granted only to aesthetic values.

The problem that Guattari had proposed in the chapter of *Chaosmosis* devoted to the aesthetic paradigm has to be revisited within this new framework.

*****

I have always been struck by the way in which Félix approached works of art and artists. Félix followed a great number of artists through their different efforts: Enzo Corman, François Pain, Gerard Fromanger, Jean-Jacques Lebel, Jean-François Léaud, Annie Raati, Tonino Guerra, Laura Betti, and many others, more or less noteworthy.[2] He was someone who was casual in the way that he looked at a work of art because he did not distinguish the work from the flow of existential events in which it was situated. Only at times did he instead stop in a slow suspension, a sort of meditative sinking, before an artwork.

I remember the visit to the Jean Tinguely exhibition held at the Beaubourg: Félix sat down on the ground at a certain point, in one of the rooms, and for a while, I had the impression that he had fallen asleep. He had not done so, but rather had let himself become drowsy by the cogwheels and the whirling rhythm. The aura was definitively forgotten in this way of getting closer to artists' activities. What had the most value for him instead were the environment, daily life, urban space and the way in which the artistic gesture put into movement new possibilities of existential assemblages within this environment, this daily life, this urban space. Félix took as given reality the becoming true of the Dadaist revolution, its definitive realization in daily life. But in his attitude there was none of this somewhat snobbish and nostalgic irritation against the effects of banalization caused by that generalized aestheticization that intellectuals often have. The banalization, the diffusion, the popularization were a given, they were taken as such and were no longer an object of discussion, and especially not a source of scandal. For Félix there was no nostalgia for the aura. Rather, the problem of art for him was completely located in the possibility of putting in motion assemblages of enunciation, housing, urbanism and technologies.

*****

The aesthetic paradigm in *Chaosmosis* comes to assume an essential role in redefining the historical and social perspective, and it is fully integrated into the problem of ecosophy.

Ecosophy, an environmental consciousness adequate to the technological complexity of late modernity, is based on the decisive character of aesthetics in the prospect of ecology.[3]

Aesthetics is the science that studies the contact between the derma and different chemical, physical, electromagnetic, electronic and informational flows. Hence aesthetics has much to do with the modern psychopathology of contact, and also with its psychotherapy.

*****

The rhizomatic vision conceptualizes the universe as a continuum of diverse, irreducible and yet juxtaposed entities in epidermal contact with each other. It is both an organic and inorganic continuum, animal and machinic, mental and electronic. One event does not approach another one through logical reasoning, nor through historical consequentiality, or through dialectical attraction-repulsion.

There is no necessary consequentiality in juxtaposing one event with another, nor any implicit logic in juxtaposing one sign with the next, nor any necessity for isomorphism between a composition and the world.

In any composition there is only the pleasure of the conjunction, me and you, this and that, the wasp and the orchid. The part is not completed thanks to conjunction with the part, nor does placing one next to the other cause any totality to arise. The conjunction is the pleasure of becoming other. From this arises the adventure of knowing, the adventure of erotic pleasure and of artistic creation.

> A rhizome has no beginning or end; it is always in the middle, between things, interbeing, *intermezzo*. The tree is filiation, but the rhizome is alliance, uniquely alliance. The tree imposes the verb 'to be', but the fabric of the rhizome is the conjunction, 'and ... and ... and ...'. This conjunction carries enough force to shake and uproot the verb 'to be'. ... To establish a logic of the AND, overthrow ontology, do away with foundations, nullify endings and beginnings. (Deleuze and Guattari, *A Thousand Plateaus*, 25)

There is nothing other than conjunctions, conjunctions of signs and corpuscles.

Signs acquire meaning in conjunctions and corpuscles acquire form through conjunctions. One mode of conjunction is sense, and another mode is pleasure.

Art seeks its contact point in sense and pleasure.

Part II

6
User's Manual

User's manual for Hegel

Don't misunderstand me, it has never been my intention to come out in support of improvisation. It's true, I detest those who spend a great part of their lives dissecting cadavers of thought. But I do not want to preach philosophical approximation either.

Quite to the contrary, I would like to approach philosophers with that broadminded lightness that I learned from rhizomatic style – as a tool box, as Deleuze and Guattari said about philosophical concepts.

Open the box, rummage around nervously, put aside the pieces that are not useful, take what might be useful to complete a passage, to make fear pass by, to open a window, and then move along with the disjointed step of a schizo craftsman.

> Our dialectics weren't derived from Hegel's cunning.
> Through the battle din it burst into our verse
> When bullets from our guns sent bosses running
> The same as we had run from theirs at first. ('Aloud and Straight', in Mayakovsky, *Selected Works*, 284)

My generation took part in the final battle about which Vladimir Mayakovsky spoke.

Because we did not learn the dialectic from Hegel. We spoke Hegel because there was no other way of relating the story in which we were then living – and we said: one divides into two, and in this unity there is a struggle, and without struggle no unity emerges – because it was the simplest way of speaking about what was happening. We believed that the infinite complexity of events in the world could be reduced to a single process. Then at a certain point we understood that the dialectic was a logical trap that little by little had ensnared us and prevented us from seeing the world, the swarm of events and persons, of solitudes and encounters without meaning, without loyalty, and without foreseeable conclusions.

Thirty years after his renaissance, Hegel seemed buried. He still is, in a certain sense, and we can affirm without fear of rebuttal that nobody will ever be able to exhume his body.

And yet (a miracle of this perfectly circular thinking) it seems impossible to move beyond it, because that thinking wished itself to be and was the thought of overcoming (*Aufhebung*). This is the essential point for a definition of modernity, which is a world of forms in constant overcoming, and thus appears impossible to be overcome.

This is what I say when I say Hegel: the totalizing thought of overcoming that both sweeps away and maintains.

Today, I know, Hegel seems to be the distant residue of an era in which one could think in terms of totality and totalization through overcoming. Today we speak rather of globalization, of globality. And we cannot think of two more diverse, more distant concepts (not antithetical, mind you, but simply other) than these: globality and totality.

The concept of totality implies a process in which the parts disappear and at the same time are maintained, they are negated in order to be recognized as a simple moment of a higher synthesis in which the parts are both negated and maintained. We are dealing with an essentially historical concept. Totalization is the process through which whatever is given as separate comes to be dissolved and subsumed, that is, translated into its negation, thanks to which it finds its spirit, its vocation, and its truth.

The concept of globality is rather a functional concept that implies integration, intersecting, combinations and interfaces. These have nothing to do with negation or with overcoming. Parts remain what they are, and nobody cares what your spirit is made of. What is important is that you can interface your surface with the surface of another entity, of another part. Globalization means functional integration of parts that maintain their independent partiality and nevertheless function together, thanks to interfaces that do not change their interiority, but connect by recombining together. In globalization, we participate in the connections between discreet and recombined elements, while in Hegelian totalization, we observe the work of negation accomplished by a continuous ensemble from which a new ensemble continuously emerges. For me, to encounter Deleuze and Guattari's thought, to acquire their way of employing concepts and words, has meant undertaking a long (incalculable and interminable) march of liberation from Hegel.

Thus, I apologize to the young reader if my language might at times appear unclear. It is not his or her fault, but mine. And it is not only a question of language. The lack of clarity comes from the fact that I carry in my discourse the trace of a style of thinking that has no vitality, that is untranslatable into the daily experience of post-historical existence. And yet, it is from this style that the tangled multi-plane routes of Deleuze and Guattari's thought depart through rhizomatic proliferation.

User's manual for Heidegger

I have no affinity for Heidegger. His human and political experience is that of a cowardly and reactionary petty bourgeois, the little man who, out of love for tranquillity, allowed Hitler to do what he wanted. In his works, however, there are glimmers of philosophical genius. Dim glimmers, glimmers that darkly colour the late-modern historical horizon.

His most beautiful book, in my view, is *Off the Beaten Track* (*Holzwege*). The paths about which he talks are those that lead to the heart of the woods. But these paths never join in any centre, because there is no heart of the woods, no centre in the woods. The dialectical hope is thus cancelled, and no possibility of overcoming is available.

Heidegger's thought proclaims the end of modern hope. But at the same time, it delineates the ontological architecture of a post-historical and post-dialectical universe.

This architecture can be understood according to two concepts that anticipate the formation of the digital techno-sphere, the concept of the 'language of technology' and the concept of 'the era of the world image'.

When Heidegger speaks of the language of technology, he does not mean that technology is an object of language, that language 'states' the technical like one of several possible discursive elements. It means that technology is constituted as a language, as an access to meaning. With Heidegger we understand that human civilization has moved across the threshold of a universe (the techno-sphere) in which forms of automation take the upper hand over choices, possibilities of alternatives, and historical dialectics.

And when he speaks of the era of the world image, we likewise must not understand that in the modern era the world can be represented (with techniques of photographic, cinematographic or televisual reproduction). We must understand something more radical. The world no longer exists except as an image. The gesture of conceptual creation accomplished by Heidegger is thus precious and of the greatest importance, if we wish to understand how philosophy has evolved over the last decades, the decades in which Heidegger has been read and interpreted. But his is also a repulsive gesture because it removes any breath of freedom and any generosity from human action. Not on the ontological plane, but on the ethical plane is it possible and necessary to free oneself from the depressing Heideggerian conceptualization.

User's manual for Bergson

Henri Bergson's contribution to thinking in the twentieth century, it seems to me, is largely undervalued. Before Guattari and Deleuze, Bergson's message had not entirely been understood. Instead it had been taken as a kind of messy irrationalism, a marginal perspective, purely transgressive with respect to the history of Western metaphysics. When I read Deleuze, and especially when I tried to understand the Guattarian philosophical gesture, I realized that it is with Bergson that the body for the first time claims its own foundational value.

With Bergson, corporeality acquires an essential role in the ontological sphere. There is no longer a discourse of being, but finally a discourse of intensive modalities that make possible a projection of being. In the history of Western metaphysics, there has never been anyone (except perhaps Spinoza) who has stated with such clarity that life is the register of being, and that intensive duration is the register of time.

As he wrote in *Creative Evolution*, 'wherever anything lives, there is, open somewhere, a register in which time is being inscribed' (16).

Time cannot be understood from the point of view of being, but only from the point of view of vital intensity. And the rhythm of temporality is located in the organism; it's in the flow, in decline that meaning is found. Bergson's thought can with every right be defined creationism, and only in this sense, it is a vitalism: not in a Romantic, irrationalist sense, but in a very precise sense, the creation of being as projection of the lived.

7
Deleuze and the Rhizomatic Machine

After Gilles Deleuze's death in November 1995,[1] rhizomatic thought received a kind of attention that it had never previously obtained. Even the academic world seemed to become aware of the importance this thought had in the philosophical scene of the late twentieth century.

However, the academics have done a kind of courtly extraction.

Gilles Deleuze has been welcomed into the nice reception room of university respectability, while Félix Guattari remains outside. He was not an academic and he associated with bad company so that in the literature devoted to rhizomatic thought, the name of Gilles Deleuze tends to be cited while Félix's name is more or less deliberately forgotten. It matters little to me, and I lack the necessary titles to promote Felix's reception by university professors. However, beyond the details of politeness and bibliographical precision, there is also an essential philosophical question.

Let's be careful now: there is a Deleuze without Guattari, and a Guattari without Deleuze, and then there is the rhizomatic machine put in motion by the encounter between the two. And according to one of their statements that should not be taken lightly, since in each one of them there were already several people, the authors of these books were truly a crowd (cf. Deleuze and Guattari, *A Thousand Plateaus*, 3).

In any case, if you want to understand the rhizomatic machine, you cannot underestimate Félix's specific contribution.

Deleuze without Guattari moved through the entire history of Western philosophy in order to free the concept of the event from any metaphysical reduction. The irresponsible event, the light dance dear to his beloved Nietzsche, is Deleuze's contribution to the rhizomatic machine.

Guattari without Deleuze constructed a philosophical style from his psychiatric practice, from his work as a political militant, and from his training in biology and pharmacology. Out of these resulted the molecular method of the 'cut up', of montage, of decomposition and recomposition, of combinatory creation. To the rhizomatic machine Guattari brought the concrete micromaterial of his inquiry.

The crystalline acuity of the Deleuzian philosophical razor combined with the Guattarian material swarm of bio-informational principles: this is the rhizomatic machine. If you leave out a piece of it, you cannot understand how it works.

Deleuze without Guattari

If we wish to consider Deleuze without Guattari, we must first ask ourselves how to enter into Deleuzian thought.

Deleuze thought is not a labyrinth. The labyrinth unfolds in a monoplanar space; in the labyrinth, one follows a route that might lead nowhere, but nonetheless one's feet rest on solid ground. The relationship between high and low is guaranteed, at the very least. The labyrinth's plane is not traversed by any other plane. Hence we cannot escape. There are no lines of flight on a mono-planar territory.

The territory that we call Deleuze is not a labyrinth, but rather a multi-planar territory.

One walks along the streets in a labyrinth but at a certain point, you are walking in another labyrinth, on another plane, as if in an Escher drawing.

How is it possible that language produces meaning? Deleuze asked. Language does not exactly function according to biunivocal responses or monoplanar references. Monoplanarity belongs to formal languages, the languages that conventionally reduce the semantic spectrum, eliminating the indefinite elements that come from the pragmatic context, from situations. But when language possesses a body, its mode of functioning is polysemic, ambiguous, enriched by the asperity of the nonverbal. And so how is it possible to communicate? In communication, there is always something more and something less than a simple semantic transfer.

We know that any sign refers to a signified that, in its turn, is the sign of other signifieds and so forth to infinity. So the process of interpreting signs should be an infinite process. But why, then, when I sense an imminent danger and I scream, 'Careful, run away!', do the people who are with me start to run?

The question cannot be answered according to a syntactic or even a semantic logic. Language refers also to its context, to the body, to the before and after, to the imaginary and expectations, to desire. The production of sense depends on everything that serves as a context to the enunciation and to the intentionally produced signs; it depends on the fact that language crosses into and connects with other planes, proliferating in this encounter. We can attempt to describe the various planes separately, but they only function in an instantaneous manner. When I start running while screaming 'run away!' people run with me not only because of the words I have said but from the way in which I said them, from the fact that I am terrorized and that I am fleeing myself.

In communication there is a dimension at once pragmatic, affective and emotional that shares several traits with contagion. The production of sense has much to do with contagion, not only with syntactic orthodoxy and semantic pertinence. What the production of sense puts into motion are 'lines of flight' toward other planes of expression.

When two people argue, the problem is not that one is right and the other is wrong, but that they inhabit two unequal planes: the one walks the length of a labyrinth different from the other's, because the two labyrinths intersect along a line that allows flight onto the other plane, but otherwise they do not coincide. This is why Deleuze said he did not like the method of disputation. To dispute someone's thought is not very interesting, nor even very useful, because there is nothing to dispute. Rather, it is a matter of understanding on what plane a certain speaking subject is moving, and what assemblages set his process of enunciation in motion.

In fact, there was a curious aspect of Félix's and Gilles's mode of arguing that seemed particularly remarkable in situations of collective discussion. There was a kind of lightness, an excess of freedom that could be viewed as superficiality

I remember one spring day in the early 1980s, when I had to prepare a document to launch a day of struggle against the ongoing repression in Italy. There was me, Félix, and Franco Piperno seated in a garden reading texts that each of us had written separately so that we could then fuse them into a single text. I had recently seen a movie, *The Man Who Fell to Earth*, with David Bowie coming from a distant planet and remaining trapped on this world. I liked the film immensely and what particularly moved me was the idea of belonging to a race of visitors that happened to come here by chance and then were caught on planet Earth. I wrote something on this theme, about David Bowie and extraterrestrial visitors.

I remember Franco Piperno's perplexity, but I especially remember that Félix said something ambiguous and showed his approval, saying hastily something like: 'let him be, this is his trip'. At the time I was perplexed. It seemed to me to be almost a kind of psychiatrist's tolerance for the strange manifestations of some lunatic. But in reality Félix often proceeded in this way, seizing immediately an opening, a phrase, an image, to insert it into a sometimes distant, heterogeneous flow. There's no need to agree about a phrase or a word. All this quibbling about the letter of a text makes no sense, all these disputes about a concept, about a logical connection.

Any subject of enunciation follows a logic that cannot be judged, much less disputed. All that we can do with any statement is to accompany the enunciator, place ourselves on the same path as him or her, go in the same direction if we feel like it, or abandon him or her at the first turn and pursue our own route.

Sense, style

In a 1988 interview, 'On Philosophy', with Raymond Bellour and François Ewald, Gilles Deleuze spoke about the way he understood philosophy:

> Great philosophers are great stylists too. Style in philosophy is the movement of concepts. This movement's only present, of course, in the sentences, but the sole point of the sentences is to give it life, a life of its own. Style is a set of variations in language, a modulation, and a straining of one's whole language toward something outside it. Philosophy's like a novel; you have to ask 'What's going to happen?', 'What's happened?. Except that the characters are concepts, and the setting, the scenes, are space-times. (*Negotiations*, 140–1)

Style is the relation between language and its outside; or rather it is the way that the outside makes itself language. The environment, the body, clothing, the mask, the game, the relationship, seduction, power: through this system of dissymmetrical yet intersecting planes, sense becomes manifest, is placed in motion. We can call style the particular rhythm with which this whole machine of dissymmetrical planes moves: the gesture of enunciating, projecting worlds. Sense is in this style, in this reaching toward the outside. Sense, in effect, is the act of moving in a certain direction.

In *The Logic of Sense*, Deleuze wrote:

> One is established 'from the outset' within sense. Sense is like the sphere in which I am already established in order to enact possible denotations, and even to think their conditions. Sense is always presupposed as soon as *I* speak; I would not be able to begin without this presupposition. In other words, I never state the sense of what I am saying. (28)

The problem of sense is thus situated in the perspective of intensity, and not of structure, of situations and not constants, of singularity and not rules.

As we look at Deleuze's philosophical itinerary with a single glance, we see that it goes from identification to singularity, from repetition to the event.

******

Deleuze drew his imaginary panorama through the presentation of philosophical personae (Hume and Kant, Spinoza and Leibniz, Nietzsche and Bergson, and finally Foucault ...). We can speak precisely about an imaginary landscape because Deleuze maintained that the task of thinking is to shape one's inner landscape, to manage in a certain sense the temperature of being. And thus thought constructs ways of imagining possible worlds, and constructs them through the image of thought, the image of a single thought.

The books in the first part of Deleuze's work are devoted to philosophical personae that allow the delineation of the displacements and reconstructions, the openings and the cuts that Deleuze made in the living body of the modern philosophical imaginary.

Empiricism and Subjectivity, published in 1953, was a book on David Hume. Nine years later, with *Nietzsche and Philosophy*, Deleuze offered a splendidly clear reading of Nietzsche's works. In 1963, he published *Kant's Critical Philosophy*, then in 1964 *Proust and Signs*. In 1966, he published *Bergsonism*, in 1967 *The Presentation of Sacher-Masoch (Masochism. Coldness and Cruelty)*, and in 1968, *Expressionism in Philosophy: Spinoza*.

1968 is an important year for Deleuze's philosophical itinerary, a rupture that bore its fruit through the meeting with Guattari. In the first period of his work, Deleuze confronted the fundamental nodal points of modern philosophy and thus traced in this way his original space, but never through suppression or overcoming. Rather, he pursued a kind of overlapping and weaving of discursive planes, intended quite truly as imaginary planes, as prospective planes, as visions. The philosophers' words, their books, were considered by Deleuze as tool boxes from which one takes what is useful, what can be used with other pieces. But tool boxes to construct what?

To construct territories, we can say, to construct planes of consistency, conceptual landscapes along which one can climb in order to project new worlds.

Precisely in 1968, *Difference and Repetition* appeared, in which the encounter with Nietzsche returns, but the nodal concept of difference is elaborated through an enlargement of the horizon toward Hegel and Heidegger.

The Logic of Sense appeared in 1969.

Then came the encounter with Félix Guattari and the writing of *Anti-Oedipus*: all that followed can be considered the meticulous and proliferating composition of a geographical map for the future, an upward movement that allowed the flight beyond the rusted and tainted horizons of the twentieth century.

The subject and the mind

'It is the brain that says *I*, but *I* is an other,' we read in *What Is Philosophy?* (211), that Deleuze wrote with Guattari. On this topic, Deleuze's thought had been prepared at length through a long reflection on experience and the subject.

At the end of the opening paragraph of *Empiricism and Subjectivity*, Deleuze constructs an idea of empiricism based on the claim of the irreducibility of experience:

> Hume's project entails the *substitution of a psychology of the mind by a psychology of the mind's affections*. The constitution of a psychology of the mind is not at all possible, since this psychology cannot find in its object the required constancy or universality; only a psychology of affections will be capable of constituting the true science of humanity. (21, emphasis in the original)

It is necessary to substitute a psychology of the mind by a psychology of its affections, says Deleuze. If we wish to study the mind, we will never find anything constant and necessary. In this way, with a decisive gesture, suddenly all of structuralism and all mentalist reductions are swept away. The empiricism of which Deleuze speaks is an empiricism of experience, not of something dead, but an empiricism of the experiential event that punctually, singularly and uniquely influences the act of cognition. This act is singular, it happens here and now, within determinate existential conditions, in the presence of an unrepeatable object. This is even a distinctive trait of Deleuzian style: a passion for the singular.

In his book devoted to Leibniz [*The Fold*], published in 1988, Deleuze says what he considers to be the most passionate problem for philosophy: 'Not how to attain eternity, but in what conditions does the objective world allow for a subjective production of novelty, that is, of creation?' (79).

The event is the occurrence *qua* experiential singularity. How does an event arise? Here is the principal cognitive problem, the one that Deleuze approaches in discussing empiricism. Reading David Hume's *A Treatise of Human Nature*, Deleuze undertakes the discovery of some concepts that suddenly become connected:

> The mind ... is given as a collection of ideas and not as a system. It follows that our earlier question can be expressed as follows: how does a collection become a system? The collection of ideas is called 'imagination', insofar as the collection designates not a faculty but rather an assemblage of things, in the most vague sense of the term: things are as they appear – a collection without an album, a play without a stage, a flow of perceptions. (*Empiricism and Subjectivity*, 22–3)

'We have the most distant notion of the place where these scenes are represented, or of the materials of which [the place] is composed,' Hume says in the *Treatise* (253). And Deleuze observes:

> The place is not different from what takes place in it; the representation does not take place in a subject. Then again the question may be: *how does the mind become a subject?* How does the imagination become a faculty? (*Empiricism and Subjectivity*, 23, emphasis in the original)

Representation does not reside in a subject in as much there is no subject without there being mental activity or a cartography of the world: memory, imagination, perception, representation, projection. All of this activity does not belong to the subject; it *is* the subject.

How does a collection become a system? The systematic character of the mind, of mental activity, resides nowhere if not in the imagination, in the attribution of sense to a certain landscape, to a certain mode of functioning, to a certain rule.

This is where the problem of the subject – or rather the problem of subjectivation – arises. What is a subject? asks Deleuze. And he suddenly discards the transcendental presuppositions of the Kantian derivation, and even the historicism typical of the Hegelian tradition.

In Deleuze there is no subject – this word is always greeted a bit coldly, with a certain suspicion. There is no subject, but rather subjectivation, stylistic stabilization of the mind, variations around an intention.

> The depth of the mind is indeed delirium, or – same thing from another point of view – change and indifference. (*Empiricism and Subjectivity*, 23; cf. Hume, *Treatise*, 125)

And on the other hand, ideas are linked in the mind, but not by the mind. There is no subjective principle connected to the mental: the relation is not what connects, but what is connected.

'Subjectivity is determined as an effect; it is in fact an *impression of reflection*' (*Empiricism and Subjectivity*, 26, emphasis in the original).

The subject is the rule of construction with which the collection of ideas (and the emission of enunciations) constitutes a world. The process of subjectivation is one with the work of schematization, of conceptual construction of the world. And the world is the site of psychodynamic integration of innumerable mental projections.

Does ontology then become dissolved in a game of mirrors?

Deleuze answered the question of Being ironically. Rather than speaking about ontology, for him we should talk about eventology, about the science of the unrepeatable, of the inconstant. The event is the happening of experience in its singularity.

Deleuze explores this eventology through one of his visions of empiricism, one of his definitions of the experiential given:

> What is the given? ... It is movement and change without identity or law ... 'Everything separable is distinguishable and everything distinguishable is different.' This is the prin-

> ciple of difference ... (*Empiricism and Subjectivity,* 87; Hume, *Treatise,* 18)

Thus from start to finish, Deleuze addresses difference, that is, the emergence of novelty, of the given, and not identity, constants, laws, what has permanence.

> We must begin with *this* experience, because it is *the* experience. It does not presuppose anything else and nothing else precedes it. It is not the affection of an implicated subject, nor the modification or mode of a substance. (*Empiricism and Subjectivity,* 88, emphasis in the original)

The mind is singular. To speak about the mind, we must speak about one mind, about a route through the events that make the subject: events-affections (climatic agents, as it were, of the status of the mind) and event-actions, cognitive and practical projections of the mind in the surrounding world, the movement of groping exploration and the following of maps.

Dialectic and genealogy

Deleuze's empiricism is the way through which eventology can be constructed. In *Nietzsche and Philosophy,* he faces this problem in all its breadth. In this book, Deleuze traces the philosophical landscape from which his notion of the event emerges. Nietzsche's thought constitutes the line along which he proceeds in order to move beyond the abyss that opened with the collapse, with the disappearance of God, with the disappearance of an ontological foundation of values. In this book, Deleuze says that Nietzsche's thought should be understood as an answer to the dialectic. The relationship between Nietzsche and Hegel, although apparently marginal in the text, is in reality of the greatest importance.

Contrary to what many maintain, there are many reasons to suppose that Nietzsche thoroughly knew Hegel's thought. In fact, according to Deleuze, we are bound to understand very poorly the ensemble of Nietzsche's work if we don't understand against whom his fundamental concepts are directed.

> It is sufficient to say that dialectic is a labor and empiricism an enjoyment [*jouissance*]. And who says that there is more thought in

> labor than in enjoyment? Difference is the object of a practical affirmation inseparable from essence and constitutive of existence. Nietzsche's 'yes' is opposed to the dialectical 'no'; affirmation to dialectical negation; difference to dialectical contradiction; joy, enjoyment to dialectical labor; lightness, dance, to dialectical heaviness; beautiful irresponsibility to dialectical responsibilities. (*Nietzsche and Philosophy*, 9, translation modified and corrected)

Effectively, the fortune of Nietzschean thought in the course of our century is closely linked to a process of internal crisis, to the gradual crumbling of dialectical thought. The renaissance of Nietzschean thought during the 1970s and 1980s is linked to the abandonment of a teleological perspective that in the course of the century had been associated with the historical realization of the Idea through revolutionary political movements.

The teleological perspective assumes that following the existing totality of capitalism there must be a higher totality. The fragmented totality of alienation, of contradictions, extended out toward an overcoming, toward a collapsed totality that recombines what is now presented as separate.

The real process, the becoming of men and things, is weighted down by a finalistic burden: the Idea has to be realized through work, war, and suffering. All of this is the dialectic.

Deleuze saw Nietzsche as the thinker that conceived the autonomy, the lightness of the event, the happening of the singular event at the outside of any teleological perspective in which duty and guilt posit themselves above the path of the singular.

The dialectic establishes historical finality as a tribunal to judge the truth of actions and words: values are founded in the dialectic as moments of the becoming true, of the realization of the Spirit.

Nietzsche abolished any criterion for the foundation of values, any spiritual or historical possibility of becoming true. The problem is the creation of values, and their truth is established in relation to the mutations and becomings of interpretation.

In this sense, Nietzsche substitutes genealogy for realization: the value of values is defined according to the possibility of the will that wants it, of the imagination that imagines it.

'The philosopher is a genealogist rather than a Kantian tribunal judge or a utilitarian mechanic' (*Nietzsche and Philosophy*, 2).

A genealogical thought does not pretend to abolish singularity in the name of historical truth, but instead conceives of history as a deployment of different wills: the will that projects values, constructs worlds of reference; the will to power is a hermeneutic force, a force of interpretation, the attributions of differential sense, the differential possession of the world.

'There is always a plurality of senses, a constellation, a complex of sensations but also of coexistences which make interpretation an art' (*Nietzsche and Philosophy*, 3-4).

The problem of truth is connected to the theme of the mask. If the very foundation of metaphysics is removed with Nietzsche, if metaphysics as foundation of the history of the world is corroded, and the history of the world is revealed – following the improvised conceptualization of nihilism – as a voyage without purpose and without foundation, in which the consciousness of the death of God predominates, then there remains no possibility for founding truth metaphysically.

Hence, truth must be thought in singular terms, as a gamble (and here is the departure point of modern existential reflection) and as will to power. Truth cannot be placed only in a sphere of singular subtraction, of singular creation of new values, and thus in the sphere of an affirmation of power, and not of subjugation, of belonging, of dependence. Truth and will to power are thus on the same plane, the plane that Deleuze defines, in his language, as the plane of the event's singularity and of the singular event.

The concept of difference

When one understands it correctly, the concept of the eternal return goes right to the heart of the question of the event as difference. But the notion of difference lends itself to all sorts of misleading interpretations, and the political use of this concept can result in dangerous identitarian effects.

What does difference mean? Does difference perhaps mean remaining as close to oneself as possible, protecting something that does not want to be rendered homologous or the oppositional affirmation of one's being? But if we understand it this way, difference means identity and then difference becomes a principle of identification which allows us to recognize and individuate.

On the contrary, the concept of difference really must be taken in the opposite direction, in the direction of becoming and not of being, of drift and not of identity.

And this problematic is already implicit in the Nietzschean notion of eternal return.

The reading of the concept of the eternal return given by Deleuze places itself precisely on this plane: difference or repetition, becoming other or claiming one's own identity.

> According to Nietzsche the eternal return is in no sense a thought of the identical, but rather a thought of synthesis, a thought of the absolutely different which calls for a new principle outside science. This principle is that of the reproduction of diversity as such, of the repetition of difference: the opposite of '*adiaphoria*'. (VP II 374 'There is no *adiaphoria* although we can imagine.') ... The eternal return is not the permanence of the same, the equilibrium state or the resting place of the identical. It is not the 'same' or the 'one' which comes back in the eternal return, but return is itself the one which ought to belong to diversity and to that which differs. (*Nietzsche and Philosophy,* 46, translation modified)[2]

Focusing on the Nietzschean question of the eternal return, Deleuze lays out the question of the singular event, that is, of difference understood as singularity. Singularity of the event and the generation of identity are the two poles of the thinkable.

Is the task of thought to understand the conditions of possibility of the unrepeatable, or is it to bring together the rules of repetition? This is the question that Deleuze sees at the heart of modern Western philosophy.

Within Hegelian thought, the polarity of the singular event and of the generation of the identical finds a complete systematization. For Hegel, experience as singular event is an appearance that needs to be dispelled.

The place for the dialectical affirmation of Truth is History, and in history truth makes itself Subject, abolishing and overcoming the singular immediacy of the event. Only thanks to this negation and abolition does identity come into being as a result. As Hegel wrote, in *The Phenomenology of Mind*, 'the real subject-matter is not exhausted in its purpose, but in working the matter out' (69).

The thing in itself [the real subject-matter] does not have truth in its immediate difference, but rather in the deployment of its identity, of its being identical through the cognitive and practical mediation of the Subject.

> The distinctive difference of anything is rather the boundary, the limit, of the subject; it is found at that point where the subject-matter stops, or it is what this subject-matter is *not.* (*The Phenomenology of Mind,* 69, emphasis in the original)

Hegel's philosophical project is indeed stated here: to explicitly abolish difference affirmatively in order to introduce identity, not through a formalist, intellectualized process of substituting the concept for the thing, but rather through a process of negating the thing, of subsuming the singularity of the thing in the becoming that the work of the negative deploys in history.

This is what Hegel declares:

> Here is what I have wanted to do: to collaborate in bringing philosophy closer to the scientific form, closer to the point where it will be able to shed the name of '*love* of knowledge' and to become 'true *knowledge*'. (*The Phenomenology of Mind,* 70, emphasis in the original; translation modified)

What did Hegel want to do? He wanted his philosophical project not to be purely intellectual, since it is not aimed at knowing the identical, but at creating it through the abolition of difference, and not in seeking to know the truth, but by bringing it into being through the abolition of singularity.

The abolition of difference is not a purely intellectual process, nor does it consist in going back from the particular to the general following a purely formal method. The abolition of difference, in Hegel, is an historical process, not merely an intellectual one.

History is the process within which the singularity of the event, its difference, comes to be annulled, or rather absorbed, and runs into its other, into the ideal identity, through the work of negation and mediation of the mind. Reality thus loses its lightness; the event becomes solidified as historical-dialectical Necessity.

Hegel creates the paradigm of historical legitimation, and thus founds in a powerful way the totalitarianism of necessity that dominates the political experience of modernity. Hegel identifies the dialectics as the modern form of Fate. *Fatum,* that is, what is said, the verdict that has been pronounced, and that consciousness must rediscover in the process of its historical realization.

The method of history is the method of this necessity that abolishes singularity and experience.

*****

To this method – we know all too well – the suffering singularity refuses to submit (albeit desperately, since modernity does not leave the single any way to escape); furthermore, questioning the past about the innumerable possibilities that it has annulled and suppressed is the complete opposite of the historical method and of politics. Modernity turns on a machine to reduce the diversity of human existence, or more precisely a machine producing a kind of homologization that idealistic philosophy assumes as perfect, while in reality modern history shows it to be interminable and totalitarian. The totality is never realized as the becoming true of particularity, but limits itself to suppressing particularities by overwhelming them with a policing machine.

The defeated, the deported, the exiled, the persecuted, the exploited, the disturbed all rebel against this reduction and this idea of necessity. This historical necessity is the sanction of an intolerable suffering, of the humiliation and annihilation of singular life.

How is it possible to tell the story of all that hasn't happened in history? How is it possible to recount all of these nullified events, all of these negated existences, all the forms of life and consciousness that never reached completion because they were destroyed and eliminated by historical necessity, by the force of what had to happen just because of the hypostasis and divinization of economic laws?

The living singularity is suppressed to make room for history. The experience of the living and conscious organism is annulled in order to make a path for the Subject.

> The living substance is that being which is truly subject, or what is the same thing, is truly realized and actual solely in the process of positing itself, or in mediating with its own self its transitions from one state or position to the opposite. As subject it is pure and simple negativity ... True reality is merely this process of reinstating self-identity, of reflecting into its own self in and from its other, and is not an original and primal unity as such. It is the process of its own becoming, the circle which presupposes its end as its purpose, and

> has its end for its beginning; it becomes concrete and actual only by being carried out, and by the end it involves. (*The Phenomenology of Mind*, 80–1)

Thus, Hegel first said that difference is not the truth of the thing, but rather its limit. Now he is saying that the truth of the thing is in the overcoming of its reality, of its experience: the truth of the thing is in its dissolution in the historicity of which the Subject (not the living and conscious organism, but the Subject) is the relay and the mediator.

In Hegel, the beginning is the end. Experience is not the experience of anything, but only the return to truth, to necessity, of duty, because truth, necessity and duty are freed (even if in a violent and murderous way) from the singularity of experience. In Hegel, 'experience' is nothing other than retracing the necessary path that brings the singular to become Subject. Experience is the dialectical dependence on history. Hegel says that no one is allowed to flee from history if one wants to have experience. Experience is created in this dependence; this after all is the essence of the modern: the subsumption of every molecule of lived time inside the general category of historicity. In this sense the modern is the ruthless historical colonization – economic, communicative, technical – of human time, of our mental time and activities. Only what has been rendered homologous by the historical work of negativity approaches the Hegelian definition of experience.

> Hegel himself identifies experience with the dialectic ... The system is not to be conceived in advance, abstractly; it is not to be an all-encompassing schema. Instead, it is supposed to be the effective center of force latent in the individual moments. They are supposed to crystallize, on their own and by virtue of their motion and direction, into a whole that does not exist outside of its particular determinations ... Experience is supposed to be something immediately present, immediately given, free, as it were, of any admixture of thought and therefore indubitable. Hegel's philosophy, however, challenges this concept of immediacy, and with it, the customary concept of experience. (Adorno, *Hegel: Three Studies*, 56–7)

The Hegelian critique of immediacy acts like a kind of defoliant, a kind of ruthless annihilation of anything that lives. Nothing that flees the incessant work of mediation (labour, technique, communication, civilization) can be thought and experienced. In this way, the singular is cancelled and persecuted until it finally gives way to historical subsumption, until it is recognized as an economic, social and historical agent, thus abandoning its independence, its singularity, in order to be admitted to socialization, inscription and finally becoming a subject.

Experience in Hegel is therefore the revelation of historical destiny. Faced with the violent revelation of necessity, singularity has to bow to this knowledge. In Hegel the reduction of experience to knowledge is implemented through the historical process intended as the subsumption of historical individuality within the productive, economic, political and state collectivity.

This is how modernity develops, as the domination of social totality over lived time.

*****

Hegel's place in the history of Western thought is so big and essential because it is with Hegel that Western philosophy conceptualizes a process of annihilation of experience and the imposition of the historical dialectic on a humanity without history.

The imposition of integral historicity coincides with the start of a process of colonization (interminable, but implacable) that is the colonizing process of experience by exchange value and money, by a unifying principle, by a unidirectional and conventionalizing temporality, thus by capital and by its dialectical abstraction, history.

Modernity begins when the historical reduction of becoming is solidly installed in the heart of lived experience, corroding and causing it to disappear from within and emptying it completely into the simulacrum. Becoming, now made historical, acts as an active annihilation and subsumption of difference, as the colonization and homologization of singularity by means of a pervasive and totalizing principle.

The Hegelian system is constructed as a perfect universe that cannot be upset or overcome because it assumes and implies contradiction and overcoming, making of them its spirit, its sustenance and the very sense of its process. It is said that the Hegelian system creates a system of circles that encompass each others, and trying to overturn a circle is a vain undertaking. The only gesture that can free thought from the prospective game of Hegelianism is the gesture of abandonment, of dissolution in laughter.

Georges Bataille recognizes this liberating power of laughter in *Inner Experience*.

And from Bataille's laughter emerges Jacques Derrida's deconstruction, who in 'From Restricted to General Economy: A Hegelianism without Reserve' in *Writing and Difference*, wrote the following:

> To laugh at philosophy (at Hegelianism) – such, in effect, is the form of the awakening – henceforth calls for an entire 'discipline', an entire 'method of meditation', that acknowledges the philosopher's byways, understands his techniques, makes use of his ruses, manipulates his cards, lets him deploy his strategy, appropriates his texts. (252)

The concept of difference – as it is elaborated by Deleuze – cannot be understood without referring to the whole development of its discourse; that is, it cannot be understood without reference to the concept of deterritorialization, of desire. Deleuzian difference, in fact, is not the positing of an oppositional alterity differentiating itself from something else: in the final analysis that would mean identity. Difference is the process of becoming different from oneself.

*****

To understand the formation of the concept of difference throughout Deleuzian thought it does not suffice to refer to Nietzsche and to the lineage of German thought that confronts the themes of ontology, metaphysics and history. Rather, our visual field must extend to understanding a lineage that can be associated rather with scientific thought, to organicism and vitalism: in short, to Henri Bergson. In fact, if we limit ourselves to considering Nietzsche, we do not succeed in understanding the genesis of energetic and organicist concepts such as desire, or flow, that are essential for seizing the specifically Deleuzian curvature of the concept of difference.

> Deleuze's polemic against psychoanalysis and Marxism is again a variation of the interpretation of difference as a purification of the simulacrum from all structural reference. What has happened to difference here? Is there any sense in which we can still apply the phrase 'philosophy of difference' to a philosophy of the simulacrum that glorifies the duplication operated by a libido with no canalization or code on a 'body without organs'? Perhaps so, but only on condition that difference becomes a question of pure energy assimilable to that inequality of forces which creates the possibility

> of fluxes, the exchange processes of life. This is in fact the outcome of Deleuze's thought, and it is strongly suggestive of a return to Bergsonian vitalism, a constant source of inspiration for his work. (Gianni Vattimo, *The Adventure of Difference*, 145–6)

The complex ontology of Bergson

> The Search for lost time is in fact a search for truth. If called a search for lost time, it is only to the degree that truth has an essential relation to time. In love as much as in nature or art, it is not pleasure but truth that matters. Or rather we have only the pleasures and joys that correspond to the discovery of what is true. The jealous man experiences a tiny thrill of joy when he can decipher one of the beloved's lies, like an interpreter who succeeds in translating a complicated text, even if the translation offers him personally a disagreeable and painful piece of information. Again we must understand how Proust defines his own search for truth, how he contrasts it with other kinds of search – scientific or philosophic. (Deleuze, *Proust and Signs*, 15)

Reading Proust, Deleuze slowly brings to light a new thematic field, a new point of view: the relation between intensity and truth. In the *Recherche*, what is in question essentially is intensity, variations of perception, of memory, of pleasure. And the sphere of this reconstruction of truth is the sphere of time, which unfolds through the flow of memory.

> Time wasted, lost time – but also time regained, recovered time. To each kind of sign there doubtless corresponds a privileged line of time. The worldly signs imply chiefly a wasted time; the signs of love envelop especially a time lost. The sensuous signs often afford us the means of regaining time, restore it to us at the heart of time lost. The signs of art, finally, give us a time regained, an original absolute time that includes all the others. But if each has its privileged temporal dimension, each also straddles the other lines and participates in the other dimensions of time. (*Proust and Signs*, 24)

Already in his study of Hume, we saw that, for Deleuze, empiricism is to be understood as a theory of knowledge that places itself in the perspective of experience, of imagination, of the learning process, of remembering, from the viewpoint of experiential intensity rather than that of the object.

Now we see, through the reading of Proust, that the dimension in which experience occurs is the experience of time, and that knowledge is inseparable from it.

On the other hand, the position in which the subject of experience posits itself is that of interpretation: the signs come to be interpreted through the singularity of recollection, through the singular unfolding of intensity:

> One must be endowed for the signs, ready to encounter them, one must open oneself to their violence. The intelligence always comes after; it is good when it comes after; it is good only when it comes after ... *There is no Logos; there are only hieroglyphs* ... The essences are at once the thing to be translated and the translation itself, the sign and the meaning. They are involved in the sign in order to force us to think; they develop the meaning in order to be necessarily conceived. The hieroglyph is everywhere; its double symbol is the accident of the encounter and the necessity of thought: 'fortuitous and inevitable.' (*Proust and Signs*, 101–2, emphasis in the original)[3]

Duration, memory, vital impetus (*élan vital*) are central concepts of Henri Bergson's thought in the Deleuzian perspective. Bergson did not limit himself to delineating a psychology, but established the bases for a 'complex ontology', an ontology able to encompass multiplicity and becoming:

> Duration seemed to [Bergson] to be less and less reducible to a psychological experience and became instead the variable essence of things, providing the theme of a complex ontology. (Deleuze, *Bergsonism*, 34)

How does one manage to understand reality? Proust said: by deciphering signs through the internal instruments of intuition and emotional interpretation, following the interpretative line of memory in time. Not with an investigation of the object, but an investigation of the way in which we experience it. This is how reality is understood.

> When we are sitting on the bank of a river, the flowing of the water, the gliding of a boat or the flight of a bird, the uninterrupted murmur of our deep life, are for us three different things or a single

> one, at will ...' (Bergson, *Duration and Simultaneity*, 52, cited in *Bergsonism*, 80)

And Deleuze observes:

> Here Bergson endows attention with the power of 'apportioning without dividing', 'of being one and several'; but more profoundly, he endows duration with the power to encompass itself (*Bergsonism*, 80).

Multiplicity is the constitutive condition in which thought unfolds for the simple fact that the lived is multiple.

> Remember that Bergson opposed two types of multiplicity – actual multiplicities that are numerical and discontinuous and virtual multiplicities that are continuous and qualitative. (*Bergsonism*, 79–80)

Consciousness and intentionality are virtual, continuous and qualitative multiplicities. In the sphere of intentionality and consciousness, actual multiplicities are ordered and interpreted.

The complex ontology that Deleuze discusses, that he saw emerging in Bergson's works, rests precisely on this point: Being enters into the infinite and innumerable game of mirrors staged by time as duration. Perhaps this is why a good reader of Deleuze is also the best possible reader of Phillip K. Dick, even if Dick, I suspect, had never heard of Deleuze, who instead did indeed know him.

Reality is generated by the flow of virtual multiplicities that become actualized: consciousness, virtual multiplicity, intentionality, virtual multiplicity, which are all actualized in memory, in imagination, and in the projection and production of worlds of experience.

What Bergson called *élan vital* is the projection of worlds, the connection between the psychological and the ontological spheres.

> What does Bergson mean when he talks about *élan vital*? It is always a case of virtuality in the process of being actualized, a simplicity in the process of differentiating, a totality in the process of dividing up. (*Bergsonism*, 94)

The Deleuzian reading rescues Bergson's thought, at least partially, from the late-Romantic and vitalist context in which it is generally inserted. The Bergsonian concept of time-duration comes to be rethought in light of an ontology that becomes the background for Heidegger's thought, for the ontological perspective of *Dasein*, and the reflection contained in *Being and Time*.

Bergson conceived of time as the singular vibration of a different duration within which Being can be seen and from which it can be founded as temporality and as the development of an intensity in time.

> Psychology is now only an opening onto ontology, a springboard for an 'installation' in Being. No sooner are we installed than we perceive that Being is multiple, the very numerous duration, our own, caught between more dispersed durations and more taut, more intense durations ... Everything happens as if the universe were a tremendous Memory. (*Bergsonism*, 77)

The key to Deleuzian thought derives from this: the discourse on Being is above all a discourse of intensity since sense is the vibration and the variation of intensity that memory or imagination produces in consciousness.

> The universe is made up of modifications, disturbances, changes of tension and energy, and nothing else. Bergson does indeed speak of a plurality of *rhythms* of duration; but in this context he makes it clear – in relation to durations that are more or less slow or fast – that each duration is an absolute, and that each rhythm is itself a duration. (*Bergsonism*, 76, emphasis in the original)

Singular ontology

In 1968, *Difference and Repetition* appeared; in 1969, *Logic of Sense*, two texts that develop premises that we can find in *Nietzsche and Philosophy*. The first takes off from the question of the eternal return, the second from the problem of genealogy and interpretation as the source of sense.

> Two lines of research lie at the origin of this book: one concerns a concept of difference without negation, precisely because unless it is subordinated to the identical, difference would not extend or 'would not have to extend' as far as opposition and contradiction; the other concerns a concept of repetition in which physical,

> mechanical or bare repetitions (repetitions of the Same) would find their *raison d'être* in the more profound structures of a hidden repetition in which a 'differential' is disguised and displaced. (Deleuze, *Difference and Repetition*, xx)

Once again, as in the book on Nietzsche, the concept of difference is proposed in a framework that explicitly diverges from that of Hegel. The process of becoming is not understood in a finalistic direction; the event cannot be overcome by a totality that encompasses it – rather, the event can only be understood as untimely.

> Following Nietzsche, we discover, as more profound than time and eternity, the untimely; philosophy is neither a philosophy of history, nor a philosophy of the eternal, but untimely, always and only untimely – that is to say, 'acting counter to our time and thereby acting on our time and, let us hope, for the benefit of time to come'. (*Difference and Repetition*, xxi; citation from Nietzsche, *Untimely Meditations*, 60)

The temporal perspective within which we can understand the event is that of an uninterrupted discontinuity that cannot be totalized because it can only be represented from within.

> Eternal return cannot mean the return of the Identical because it presupposes a world (that of the will to power) in which all previous identities have been abolished and dissolved. Returning is being, but only the being of becoming. (*Difference and Repetition*, 41)

Identity can no longer be understood as the first principle, but as a principle in becoming, as the metamorphosis of the different that reflects on itself by means of a provisional process of self-identification:

> God makes the world by calculating, but his calculations never work out exactly, and this inexactitude or injustice in the result, this irreducible inequality, forms the condition of the world. The world 'happens' while God calculates; if the calculation were exact, there would be no world. The world can be regarded as a 'remainder', and the real in the world understood in terms of fractional or even incommensurable numbers. Every phenomenon refers

> to an inequality by which it is conditioned. (*Difference and Repetition*, 222)

The world is the proliferation of residues that derive from the imperfection of the divine calculations and from the interminability of historical work.

Somewhere Félix Guattari wrote that 'history is written in syncopation [*contretemps*]'. What is important in historical processes is the emergence of something that was not foreseen by laws, something that was not already regulated.

Difference, therefore, is not the non-being of something, or what is opposed to the thing, negating it from the outside. Difference rather is in the heart of the thing, is the same dynamic that permits definition and change. Definition and change, finally, are the same; if you want to define the thing, you must tell me how it changes. And the thing does not change through opposition to something else, but through the intimate imbalance that causes everything to move.

Difference is understood as a positive concept, not an oppositional concept.

If we wish to adopt the perspective of the speaker who assigns sense, that is, if we wish to adopt the point of view of sense as production, as that which comes from an origin, we must realize that difference has this prospective power and not reduce it to a dialectical pole, to a mere conceptual figure. Difference stands as the source from which sense emanates; it is the coincidence in the dice throw and requires the numerical configuration that derives from it. ('Ontology is the dice throw, the chaosmos from which the cosmos emerges': *Difference and Repetition*, 199.)

Difference is the entrance to the singular ontology around which Deleuze worked, a complex and pulsional ontology, an ontology of innumerable points of view projecting singular worlds.

The language of the event

' "What is it?" I cried out with curiosity – *which one is it?* you ought to ask. Thus spoke Dionysus, then kept quiet in his own special way, that is to say, in an enticing way,' says Nietzsche in 'The Wanderer and His Shadow', cited by Deleuze (*Nietzsche and Philosophy*, 76; Nietzsche, 'The Wanderer ...', 10).

The problem of interpretation is posed starting from the viewpoint of the will, according to Nietzschean language, from the viewpoint of temporality, according to Bergson. The two visions converge, they permeate each other: in Deleuze, Nietzsche and Bergson are viewed together.

> According to Nietzsche the question 'which one?' [*qui*] means this: what are the forces which take hold of a given thing, what is the will that possesses it? Which one is expressed, manifested and even hidden in it? We are led to essence only by the question: which one? For *essence is merely the sense and value of the thing*; essence is determined by the forces with affinity for the thing and by the will with affinity for these forces ... The question 'what is it?' is a way of establishing a sense seen from another point of view. Essence, being, is a perspectival reality and presupposes a plurality. Fundamentally it is always the question 'What is it *for me*?' (for us, for everyone that sees, etc.). (*Nietzsche and Philosophy*, 76–7, emphasis in the original)

On the question of sense, both the problematic of intention and of the will to power (in the Nietzschean sense) dominate. Here is what Lewis Carroll had Humpty Dumpty say, in a passage that Deleuze cites in *The Logic of Sense*:

> Sense resides in the beliefs (or desires) of the person who expresses herself. 'When *I* use a word,' said Humpty Dumpty, 'it means just what I choose it to mean – neither more nor less ... The question is ... which is to be master – that's all.' We have, however, seen that the order of beliefs and desires was founded on the order of the conceptual implications of signification, and that even the identity of the self which speaks, or says 'I', was guaranteed only by the permanence of certain signifieds (the concepts of God, the world ...). The I is primary and sufficient in the order of speech only insofar as it envelops significations which must be developed for themselves in the order of language (*langue*). (*The Logic of Sense*, 17–18, emphasis in the original)

With the reference to Humpty Dumpty, we enter into the world of the attribution of sense. The attribution of sense is invested in the will to power or even, if we wish, the projection of desire. Pragmatics is this dimension of signification, without which we will never be able to understand the relationship between the signifier and the signified in a text, a gesture, an utterance.

******

Here we approach, by another route, the problem of interpretation that we had already hinted at in speaking of Nietzsche. Interpretation constitutes a reconsideration of the genealogical network of attributed sense. In other terms, interpreting a sign is possible through the work of reconstructing the intention of which the sign is the bearer, the context within which this intention manifested itself, and the strategies that the speaker has actualized to obtain his or her pragmatic purpose (persuasive, dissuasive, insinuating or dissimulating, enticing or repulsive ...).

For this reason one can say that the surface is the site on which signification unfolds: the surface is the contact point among agents of sense, the exteriority that reaches toward the world of relationships, the interface between language and the world.

As long as we think that signification constitutes the depth of enunciation, we remain entrapped in the conviction that there is a kind of objective, privileged interpretation of the sign and that the unfolding of enunciation is somehow predetermined and regulated by the necessary implications of the signified. But this is not in fact the case: signification is played out in the surface relation between an intended enunciator and an intended receiver.

> The surface is the locus of *sense*: signs remain deprived of sense as long as they do not enter into the surface of organization which assures the resonance of two series. (*The Logic of Sense*, 104, emphasis in the original)

About this statement, Guattari writes:

> There is nothing less logical, less mathematical, than a language. Its 'structure' results from the petrifaction of a kind of catch-all from which its elements emerge by borrowings, amalgams, agglutinations, and misunderstandings ... There is no universality of language; nor is there any universality in language acts. Each sequence of linguistic expression is associated with a network of semiotic links of all kinds (perceptive, mimetic, gestural, image-thoughts, etc ...). Any signifying statement crystallizes a silent dance of

> intensities played out both on the social body and on the individuated body. From language to speaking in tongues [*glossolalie*], all transitions are possible. (*L'Inconscient machinique*, 25, 31, our translation)[4]

There is no longer any pretence to a normative or structural reduction, but the criterion of interpretation depends on the conditions in which the enunciation unfolds; and since enunciation follows a route that can be defined as a drift (that is, the displacement from one point to another that has no recognizable logical consequentiality), then interpretation will have characteristics that we can define as delirious (in the sense of *dé/lire*, jettisoning the rules of logical and consequential reading to opt for a reading that is as multi-planar as was the enunciation).

This is how we enter into the analytical space that takes the name of schizoanalysis. Schizoanalysis presupposes the proliferating character (one irreducible to structure) of the agents of enunciation, and further presupposes the proliferating character of the viewpoints of interpretation. Analysis therefore cannot develop by presupposing linguistic and psychological structures (of which enunciation would either be an application or a deviation). Analysis becomes schizoanalysis when it adopts the perspective of multiplicity.

If language encounters the event, then interpretation must synchronize itself with the rhythm of the enunciation. And an enunciation is no more reducible to syntactic structures than a speaker can be reduced to psychological structures. If language encounters the event, interpretation must assume as many points of view as the singularities that enter into the determination of the event.

Schizoanalaysis is this mode of interpretation that assumes the viewpoint of drift and singularity. Singularities are not entirely at one with the person, with the individual – they are rather flows of intentionality and imagination that can cohabit within a person, or can spread out along the lines of internal relations to a collectivity. The collectivity of enunciation is a singularity. That is, what defines the singularity is not the individuality of the speaker, but the belonging to the same plane of immanence of an enunciation.

The authors of *A Thousand Plateaus* speak of their collective writing in terms of a singular multiplicity: 'each of us was several persons' (Deleuze and Guattari, *A Thousand Plateaus*, 3). And this is not at all a play on words.

Folds

'What always matters is folding, unfolding, refolding,' Deleuze writes at the end of his book on Leibniz in 1988 (*The Fold*, 137). Thought is in the folding, but also in the unfolding, in the process of explication. The questions that bring together Deleuze and Leibniz and also distance them, all concern the problem of explication. The same is true for Deleuze and Spinoza.

At the centre of the Deleuzian reflection on Spinoza is the theme of expression: expression means unfolding what is folded, projecting constructively what is constituted in the interior of language in the form of complication.

> Expression is taken to be synonymous with 'emanation'. (*Expressionism in Philosophy: Spinoza*, 17)

But on the other hand, as Deleuze then writes:

> Expression is taken as another word for *explication*. (*Expressionism in Philosophy*, 17–18)

Thus at the beginning is the fold, the particular folding that being assumes within language, in the singular world of a subject of enunciation. Enunciation is the process that puts projection in motion, that is, the unfolding from which the projected world emanates, the constructed world of language, the world of linguistic interaction, the rope across the abyss of non-sense.

> Becoming expressive, that is, becoming active. (*Expressionism in Philosophy*, 288)

Deleuze returns to Spinoza at several moments, and in particular he devotes two books to the Dutch philosopher, one on expressionism [*Expressionism in Philosophy*] in 1968, and another, smaller book called *Spinoza: Practical Philosophy*, in 1981.

Spinoza restores a Stoic aura to philosophy: what is at stake in philosophy as well as in the search for sense is the happiness of existence.

Spinoza questions being from the viewpoint of affection, and in this way, he recognizes – rare among Western philosophers – the philosophical legitimacy of happiness (*felicità*).

Happiness: a word so suspect that is was banished from reflection.

And yet Spinoza puts it back in the game. Happiness is the singularity that brings its desire to fulfilment. Joyfulness is the coincidence between the singular trajectory and the refrains of recognition in the world.

These are the two nodes on which the Deleuzian reflection gazes: the first is the folding, the passion, or the affection that determines the forms of cognitive apprehension. The second is explication, that is, the expressive projection, the becoming world of language.

Two symmetrical movements, the two movements of the mind: the becoming mental of the world and the becoming world of the mind.

> Desire is 'appetite accompanied by the consciousness thereof'. It is clear from the above considerations that we do not endeavor, will, seek after or desire because we judge a thing to be good. On the contrary, we judge a thing to be good because we endeavor, will, seek after and desire it. (Spinoza, *Ethics*, Part III, Prop. 9; 109)

The notion of desire is fundamental in Deleuze's thought, and here we can see its genesis: desire is opposed to lack (it is around this question that *Anti-Oedipus* develops its entire discourse) because desire produces, realizes, creates.

And what precisely does desire create? Not so much its object in its materiality, in its causality, but the qualities of this object, the qualities of the world. Desire changes the world because it constitutes the condition of possibility for desiring projections.

Along the curve of desire unfolds the diversified action of the fold:

> Explain is a 'strong' term in Spinoza. It does not signify an operation of the intellect external to the thing, but an operation of the thing internal to the intellect. (Deleuze, *Spinoza: Practical Philosophy*, 68)

This road opened by Spinoza is where Hegel's thought starts as well: with him, the constructive power of explication is transferred from the sphere of the desiring singularity to the sphere of historical subjectivity.

But Deleuze himself wants to interrupt this line of development of modern philosophical thought, placing himself between Spinoza and Hegel, avoiding this 'strange mixture of ontology and anthropology, of atheism and theology' that for him was Hegelianism (*Nietzsche and Philosophy*, 195).

Deleuze means that it is not true that the creativity of the mind is necessarily composed of idealism, of historicism, that is, the sequel negation-realization that is the essential aspect of Hegel's thought.

If we wish to question the formation of idealism along the development line of Spinoza's practical philosophy, we need to insist on the duplicity of the perspective along which the mind-world interaction unfolds: explication presupposes implication, complication. First the world becomes fold, and then the mind is projected through the unfolding of the world.

In other words, one must closely connect the notion of fold and explication with that of the event. It is not by chance that the central chapter of the book on Leibniz, *The Fold*, is a reflection on the concept of the event.

Deleuze asks, What is an event?

> What are the conditions that make an event possible? Events are produced in a chaos, in a chaotic multiplicity, but only under the condition that a sort of screen intervenes. (*The Fold*, 76)

But what does chaos mean?

> According to a cosmological approximation, chaos would be the sum of all possibles, that is, all individual essences insofar as each tends to existence on its own account; but the screen only allows compossibles – and only the best combination of compossibles – to be sifted through ... From a psychic point of view, chaos would be a universal giddiness, the sum of all possible perceptions being infinitesimal or infinitely minute; but the screen would extract differentials that could be integrated in ordered perceptions. (*The Fold*, 76–7)[5]

The product of the selection that the screen operates on the infinite generation of possibles is precisely the event. In this way, Deleuze responds to the question posed by Leibniz, Bergson and Whitehead – the question that points to renewing the internal *modus operandi* of Western thought, of metaphysics and science: not how to attain the eternal, but rather under what conditions does the objective world allow a subjective production of novelty, that is, a creation? (cf. *The Fold*, 76).

Scientific thought undertakes the search for constants in the morphogenesis of the world. But how can we explain the unforeseeable, the irruption of the new, of mutation?

This is the central question to which Gilles Deleuze's thought offers an answer.

> I've tried in all my books to discover the nature of events; it's a philosophical concept, the only one capable of ousting the verb 'to be' and attributes. (*Negotiations*, 141)

And it is here, around the concept of the event, that we can also find the exit door from the multi-planar labyrinth-thought of Gilles Deleuze.

8
Why is *Anti-Oedipus* the Book of the '68 Movement?

People are used to saying, as if it were an established fact, that *Anti-Oedipus* is the book that brings the mark of May 1968 on to the philosophical stage. Circumstances conspire to give weight to this view: in May, Gilles and Félix had marched with the students and workers in Paris, they had met and started to build their collaborative relationship, they had participated in the anti-psychiatric adventure and in the action committees against prisons. But all of these biographical details still say nothing.[1]

Nor does it say much that *Anti-Oedipus*, published in 1972, became one of the few philosophical bestsellers and had as its prevalent readers many rebel intellectuals who were culturally formed in that period and in that movement.

One must reconstruct the evolution of Deleuze-Guattari thought in *Anti-Oedipus* (but also in *A Thousand Plateaus*, in *Kafka*, and twenty years later, in *What Is Philosophy?*) to understand in what way the conceptual weavings of modern Western philosophy are recombined therein, and thus to understand what it means precisely for Félix and Gilles's thought to express the rupture of '68 and the change of scene that this entailed.

Although being the fulfilment of the modern parable, '68 allows us to look at the twentieth century from the perspective of its dissolution, of its explosion. '68 acts as the start of a proliferation of forms of drift that are no longer reducible to any unitary history.

We can see '68 as the final chapter of the proletarian and socialist struggles of the twentieth century because it is certain that the movement of students and workers, united in struggle, was the critical fulfilment of that history.

But rhizomatic thought allows us to see '68 instead as the first chapter of a post-analytical and post-political evolution that we are today called on to continue: the anarcho-recombinatory, molecular evolution. Gilles and Félix's thought allows us to read in the sequence of the '68 and the '89 ruptures the discovery of a new historical horizon and the definitive closure of the nineteenth-century historical horizon.

In the full force of the storms that, starting in 1989, have overwhelmed the order of the Cold War which had firmly gripped the world in the vice of bipolar terror, someone said that history had ended. With this they meant that an era of peace was starting in which liberal democracy was destined to spread out across the planet without meeting any obstacles or opposition, and that the new, powerful information technologies had made power transparent.

Never was a prediction more contradicted by the facts because, as we know, the 1990s have brought on to the scene the re-emergence of nationalisms and ethnic and religious integralisms, and history has come back with its irreducible complexity and all the intricate foolishness of intention. The disappearance of any possible social alternative has brought the people to a rancorous regression while the development of the digital communication network tends to completely detach the process of formation and exercise of power from the territorial and physical dimension of society.

Nonetheless, there was a kernel of truth in this very erroneous prediction.

Between '68 and '89, the modern promise of the linear development of progress lost all foundation and credibility. The promise of modern history is a promise of universality that at a certain point was contradicted. This is why the totalizing, historical cocoon that tried to contain the intelligent life of the planet within the categories of technique and spectacle has become a kind of full body armour.

The proliferation of different forms of life and sources of communication requires breathing outside the armour of economical, political and imaginary totalization. In a word: we have to think outside the historical dimension, understood as the moderns did, a process of reduction to a single totalizing perspective.

Anti-Oedipus tried to conceptualize this tendency, abandoning the paths of historicism and telling us about the proliferations of collective agents of enunciation. In order to complete this displacement successfully, *Anti-Oedipus* rethought in an original manner the entire problematic of the unconscious. Here the contribution of Félix to the rhizomatic machine was essential. While Deleuze in his earlier work had worked on the philosophical deconstruction of subjectivity and of the mind, Guattari had worked on this same object from an analytical and political point of view. His relationship with the Lacanian school on the one hand and the group analytical practice on the other were the premises of Félix's contribution. His participation in the 1968 explosion and then the encounter with Deleuze brought Félix to a definitive rupture with Lacanian structuralism. The unconscious is not a structure, but a proliferation.

Anti-Oedipus: Machine and desire

> The traditional logic of desire is all wrong from the very outset: from the very first step that the Platonic logic of desire forces us to take, making us choose between *production* and *acquisition*. From the moment that we place desire on the side of acquisition, we make desire an idealistic (dialectical, nihilistic) conception, which causes us to look upon it as primarily a lack: a lack of an object, a lack of the real object. (Deleuze and Guattari, *Anti-Oedipus*, 25, emphasis in the original)

The unconscious discussed in *Anti-Oedipus* is a laboratory, not a theatre. The imagination overwhelms the imaginary. Deleuze and Guattari register the passage of the world under the aegis of semiosis because semiosis is a desiring projection; it is the marking of the world by any sentient organism crossing the space in which other sentient organisms have already traced their trajectories.

In the language of *Anti-Oedipus*, sentient organisms can be considered as machines. Not that there is any pretence of mechanical reduction in the Deleuze-Guattari concept of machine.

> Machines may be defined as a *system of interruptions* or breaks. These breaks should in no way be considered as a separation from reality ... Every machine, in the first place, is related to a continual material flow that it cuts into. (*Anti-Oedipus*, 36, emphasis in the original)

With the word 'machine', what is meant here is any assemblage capable of shaping reality according to its rule, a singular cut-and-stitch that functions according to its own rules.

Reality is the *cut and mix* operated by semiotic, or rather desiring agents.

Anti-Oedipus: molecular and molar

Machines and desire constitute a deep, hidden, swarming laboratory in which the flow of reality is produced at the same time as an imaginary and material secretion. The molecular dimension is that in which each semiotic agent metabolizes and assembles the real and semiotic flows that it produces.

Metabolizes and assembles, decomposes and recomposes.

The molecular is not opposed to the molar, but tells its true story.

The molar are institutions, languages, states, recognized hierarchical structures.

They are the forms of power and ideology, of consensus and representation.

Under these forms and their disputes there swarms an infinity of organic, psychic, mechanistic, cybernetic, technological segments that are in continuous recomposition.

The fragments are assembled according to principles of heterogeneity (as 'Rhizome' will later explain, the introduction to *A Thousand Plateaus* that was published eight years after *Anti-Oedipus*).

> What is the meaning of this distinction between two regions: one molecular and the other molar; one microphysical or micrological, the other statistical and gregarious? Is this anything more than a metaphor lending the unconscious a distinction grounded in physics, when we speak of an opposition between intra-atomic phenomena and the mass phenomena that operate through statistical accumulation, obeying the laws of aggregates? But in reality the unconscious belongs to the realm of physics; the body without organs and its intensities are not metaphors, but matter itself. (*Anti-Oedipus*, 283)

Anti-Oedipus: microphysics and micropsychic

In the molecular dimension we can perceive the continuity between organic and inorganic material and the becoming of the psyche as an assemblage of semiotic fragments from a brain incessantly redefining itself with respect to its neuronal material, the infinite speed of its cognitive potentiality, and even the informational flow by which it is inundated, drenched and confused.

Desiring machines are not reducible to any totalization or any foundational truth. These machines

> ... represent nothing, signify nothing, mean nothing, and are exactly what one makes of them, what is made with them, what they make in themselves. Desiring-machines work according to regimes of synthesis that have no equivalent in the large aggregates. Jacques Monod has

> defined the originality of these syntheses, from the standpoint of a molecular biology or of a 'microscopic cybernetics' without regard to the traditional opposition between mechanism and vitalism. Here the fundamental traits of synthesis are the indifferent nature of the chemical signals, the indifference to the substrate, and the indirect character of the interactions. (*Anti-Oedipus*, 288)

Deleuze and Guattari propose to look through an electronic microscope at the macro-social processes of interaction, introducing sensors into the psycho-chemical interstices, into the deep web of the Infosphere, of the becoming mind of matter-information.

The assemblages have an a-signifying character because what matters is not the support material, or the meaning of signs, but the effect that these produce in the form of assemblages: their cognitive, psychic and social effect.

The meaning of signs is secondary with respect to their functioning.

The vision of this intimacy between organic and inorganic matter, between material and psychic flows, calls to mind the words of Carl Gustav Jung, who wrote:

> Microphysics is feeling its way into the unknown side of matter, just as complex psychology is pushing forward into the unknown side of the psyche. Both lines of investigation have yielded findings which can be conceived only by means of antinomies, and both have developed concepts which display remarkable analogies. If this trend should become more pronounced in the future, the hypothesis of the unity of their subject matters would gain in probability ... The common background of microphysics and depth-psychology is as much physical as psychic and therefore neither, but rather a third thing, a neutral nature which can at most be grasped in hints since in essence it is transcendental. (Jung, *Mysterium Coniunctionis*, 538)

This Jungian intuition about continuity between the microphysics of matter and the micropsychics of the unconscious is a line that Deleuze and Guattari develop with fewer inhibitions or forms of mysticism.

> In the genetic code as in the social codes, what is termed a signifying chain is more a jargon than a language (*langage*), composed of non-signifying elements that have a meaning or an effect of signification

> only in the large aggregates that they constitute through a link drawing of elements, a partial dependence, and a superposition of relays. It is not a matter of biologizing human history, nor of anthropologizing natural history. It is a matter of showing the common participation of the social machines *and* the organic machines in the desiring-machines. At man's most basic stratum, the Id: the schizophrenic cell, the schizo molecules, their chains and their jargons. There is a whole biology of schizophrenia; molecular biology is itself schizophrenic – as is microphysics. (*Anti-Oedipus*, 289, emphasis in the original)

Truth and history in Hegel

On the plane of ideological representations, '68 conceived of itself as the realization of truth in history, according to a Hegelian kind of interpretative schema. But beyond ideological representations, this movement initiates the essentially postmodern process of liberating truth from history and of promoting truths in proliferation, or rather of autonomous and collective processes of enunciation. Little by little, from 1968 until today, we have recognized that there is no truth to be realized in History because truth does not lie in the adequacy of history to thought, nor of thought to history.

In this mutation of perspectives that was traced in the successive decades since 1968, the Heideggerian influence plays an important role. Heidegger suggested that truth is the institution of the World in the sphere of Language and of Technology. Once we realize this, there begins the long, perhaps interminable march of liberation from Hegel, as Michael Hardt showed in his book on Deleuze (*Gilles Deleuze: an Apprenticeship in Philosophy*, 1993).

*****

Deleuze's thought can be seen, on the same line as Nietzsche's and Heidegger's, as a withdrawal from the logocentric universe constructed by Hegel.

Where can one find the exits from this circle of circles of circles that is the Hegelian system? We must seek them in a field that is neither metaphysics, nor its overcoming. The exit exists in the emancipation from historical thought and from history itself, understood as totality, or rather as process of totalization.

The exit is located in the vision of a rhizomatic proliferation of singularities.

This is what Deleuze and Guattari's philosophical creationism consists of, allowing us to see beyond the horizon of the twentieth century, beyond the horizon of historical thought.

From Hegel onward, the concept of truth was withdrawn from the gnoseological field in which it had flourished in the course of the preceding history of Western philosophy and was then reintroduced in the historical field. The problem of truth for Hegel is no longer that of the correspondence of the thing to the intellect (*adaequatio rei ad intellectum*) that Hegel denounced as a purely intellectual construction.[2] That means that the problem of truth is no longer definable, that is, in purely cognitive or representational terms, but dialectically, in terms of realization, of fulfillment.

'In my view ... everything depends on grasping and expressing the ultimate truth not as Substance but as Subject as well,' Hegel wrote in the fundamental 'Preface' to *The Phenomenology of Mind* (80).

> The truth is the whole. The whole, however, is merely the essential nature reaching its completeness through the process of its own development. Of the Absolute, it must be said that it is essentially a result, that only at the end is it what it is in truth. (*The Phenomenology of Mind*, 82)

The truth is not the adaptation of the mind to the already given reality for the simple reason that reality is 'not given'. The reality that interests the dialectic is not the reality of immediacy, but that of a process of mediation, the becoming true that is only realized through the 'work of the negative', through historical, political struggle, through the transformation of the given and the realization of the idea.

This essential lesson of dynamic Hegelianism nourished the generation that prepared '68 in Europe, in Germany, in Italy, and even in France and in the USA, however differently. It is a matter of dialectical and revolutionary thinking that pulses vividly in the pages of *The Phenomenology of Mind* as much as in Hegel's later systematic works, and that was brought forth by the excavation work of authors such as Hyppolite, Kojève, Adorno and Marcuse.

The dynamic re-reading of Hegel filtered into the '68 culture through works such as *Reason and Revolution* by Herbert Marcuse:

> Reality is the constantly renewed result of existence; the conscious or unconscious process in which that which becomes something

> different from itself and identity is only the continuous negation of the inadequacy of existence; it is the subject that remains itself by becoming different. Any reality is therefore a realization, a development of subjectivity. The latter attains itself in history when its development becomes rational ... Dialectical thought began with the admission that the world is not free, that is, that man and nature exist in conditions of alienation, different from what they are ... Thought corresponds to reality only if it transforms practice itself by understanding its contradictory structure. Here the principle of the dialectic carries thought beyond the limits of philosophy. Understanding reality means understanding what things are, and this in its turn includes the non-acceptance of their belonging as raw facts. The non-acceptance, the revolt, constitutes the process of thought as well as action. (Marcuse, *Ragione e rivoluzione*, 44–5, our translation)[3]

The notion of truth here submits to an historical, dialectical twisting: the truth is no mere recognition of givenness, of being present, but is an integral process of becoming true, a process that givenness negates by putting actuality into being, which is finally the same process. Givenness, the immediacy that we can experience in daily existence, is the locus of an alienation, that is, in being other than oneself, of a discrepancy between what exists and its ideal destination, its having to be, or even its being able to be, its possibility, that will become actual only through the work of the negative. Hegel can affirm that all that is real is rational only because all that is true has passed through the work of the negative that introduces reality and brings truth to reason. The phrase 'all that is real is rational' is not the recognition of a given condition, but the vision of a process of realization that brings being to truth [*porta a essere la verità*] only insofar as it brings truth to being [*porta la verità in essere*].

*****

Truth is the purpose of history and at the same time promises to be the end of history.

Twentieth-century utopia arises from this dynamic and dialectical conception of truth as becoming true. Revolutionary movements have tried to realize this promise, and '68 can be seen as the moment in which this dream of realization found itself closest to fulfilment.

But today, as has been revealed undeniably by 1989, we know that this utopist dream accompanying twentieth-century revolutionary movements and linked to the fulfilment of '68 came to fruition as a nightmare in the century's totalitarian regimes. The idea of dialectical identity, the identity of the real and the rational, is in fact a terrifying idea that only comes about through the destruction of the life of what is different, and the annihilation of the pulsating proliferation of singularities. And in fact Adorno had foreseen this outcome.

In *Minima Moralia*, Adorno wrote: 'The whole is the false' (50), reversing the Hegelian affirmation, but staying within the dialectical sphere that becomes, in Adorno, the negative dialectic. Hegelian truth is the relation of the unity between the subject, the motor of the historical process, and the world that becomes Spirit through historical appropriation and reveals its ideal being. To remain in the sphere of Hegelian thought, to the extent that he is concerned with method and style, Adorno recognized that in history there is no overcoming, no conciliation, no totality to be realized and, we can say, no truth. Historical totality is not, as Hegel pretends, the wholeness of the true, but the wholeness of the false.

In this way, the problem of the truth is suspended in the anticipation of an answer that probably is not located inside history. The finality of realization and of overcoming remains a non-realized finality; to turn it into the core of human activity means fating it to interminable torment. Twentieth-century history is, in effect, the revelation of the fact that the tension toward a totality missed its goal, or achieved it uniquely in the form of totalitarianism.

*****

Philosophy which, with Hegel, had promised to realize itself and thence to die missed the moment of its realization. Now we must bring it back to life, but by asking it another question. Philosophy can no longer promise its death-realization in a pacified world; instead it must propose itself as the thinking of singularity in a world in which totality is definitively and inevitably the negation of the human.

Adorno said, the whole is the false, but his reversal is Hegelian, like any reversal: the dialectic of the false and the true lingers in the Hegelian conceptual domain, and little by little we need to get out of this sphere. Reversing a sphere makes no sense since we always find ourselves faced with the same sphere; the only thing to do with it is to dissolve it, to molecularize it, to draw from it a proliferation of singular slivers.

Adorno upset the notion of subject to the point of taking it to its far limit, to the point of abandoning it and of thinking the individual. We have to go farther and we have to think singularity. This is how we arrive at the recognition of Deleuze and Guattari and of *Anti-Oedipus* as the book that interprets the '68 movement.

Note well that this book does not interpret the consciousness that '68 had of itself, which was entirely internal to the Hegelian field. In fact, it interprets this dynamic that carried '68 beyond its consciousness, leading '68 to cause the explosion of the sphere in which it had developed and had been conceived and represented. *Anti-Oedipus* works out this movement by abandoning the totalitarian frame within which the twentieth-century social consciousness was determined.

'68 as refrain (*ritornello*)

Adorno had already indicated the collapse of the Hegelian edifice, adopting the perspective of the individual particularity, and not of the becoming of subjectivity.

'History does nothing, does not "possess vast wealth", does not "fight battles"! It is a man, rather, the real, living man who does all that, who does possess and fight,' wrote Marx and Engels in *The Holy Family* (cited by Adorno, *Negative Dialectics*, 304). And Adorno wrote: 'To this day history lacks any total subject, however construable. Its substrate is the functional connection of real individual subjects' (*Negative Dialectics*, 304).

Dialectical thought instead attributes to history a demiurgic power, independent of real subjects, almost as if the dialectic were the truth of these real processes, stripped of its elements of concrete individuality. Dialectical thought projects the shadow of history onto the real facts of the human mind, onto the perception of time, onto activity, onto the building of relations.

> History is equipped with those qualities because society's law of motion has for thousands of years been abstracting from its individual subjects, degrading them to mere executors, mere partners in social wealth and social struggle. The debasement was as real as the fact that on the other hand there would be nothing without individuals and their spontaneities. (*Negative Dialectics*, 304)

Adorno says here that historical totality is not the Hegelian revelation of truth, but instead the affirmation of an absolute lie.

> No universal history leads from savagery to humanitarianism, but there is one leading from the slingshot to the megaton bomb. It ends in the total menace which organized mankind posed to orga-

> nized men, in the epitome of discontinuity. It is the horror that verifies Hegel and stands him on his head. If he transfigured the totality of historic suffering into the positivity of the self-realizing absolute, the One and All that keeps rolling on to this day – with occasional breathing spells – would teleologically be the absolute of suffering. (*Negative Dialectics*, 320)

Hegel is starkly turned on his head: there is no truth at the end of the historical process, no positive totality beyond contradiction, no history beyond prehistory. The conclusion of the process of modernity is not the positive totality that '68 had thought to bring about. On the contrary, it is the deployment of an absolute lie in the form of totalitarianism, indeed of totalitarianisms. Archaic totalitarianisms of the land, of blood, of race, even the utopist ones of class, of socialism, of the party, and finally the most powerful and totalizing totalitarianism: the totalitarianism of money, of value, of the code: the totalitarianism of communication.

Anti-Oedipus: desire and territoriality

> Desire can never be deceived. Interests can be deceived, unrecognized or betrayed, but not desire. Whence Reich's cry: no, the masses were not deceived, they desired fascism, and that is what has to be explained. It happens that one desires against one's own interests ... (*Anti-Oedipus*, 257)

The process of subjectivation is constructed around the investments of desire.

The subject does not come first before desire; rather, it is desire that trails after a trace that we can call subjectivation. There is no subject that desires, but a field of assemblage in which organic, machinic, historical, sexual, etc., segments abound. Desire is the force that guides processes of assemblage [*agencement*] and sets in motion the process of subjectivation while directing it.

If I recall well, they do not yet speak about refrains [*ritournelles*] in *Anti-Oedipus*, but the concept of the refrain – that Guattari had just elaborated in his 1979 *The Machinic Unconscious* [*L'Inconscient machinique*], and then in *A Thousand Plateaus* – is essential for understanding how the process of subjectivation works in Deleuze and Guattari's thought. In fact, it is through the refrain, that is, through a singular semiotic flow capable of creating assemblages with the surrounding semiotic landscapes, that the process of subjectivation is realized. This process is concretized through refrains that allow an assemblage between a conscious organism and surrounding semiotic constellations that enter into vibration with it.

A discourse on the investments of desire must take into account the effects of deterritorialization and reterritorialization that take place around the subject, attracting it and repelling it.

> Civilized modern societies are defined by processes of decoding and deterritorialization. But *what they deterritorialize with one hand, they reterritorialize with the other*. These neoterritorialities are often artificial, residual, archaic ... (*Anti-Oedipus*, 257, emphasis in the original)

Paranoid investments of desire are determined through the encounter between a process of deterritorialization (for example, the effects of capitalist globalization on local cultures) and a movement of reterritorialization that aims at reconstructing marks of identity through stabilizing borders, through aggression and the individuation of difference, and through the criminalization of alterity.

Anti-Oedipus proposes to adopt schizophrenia as the point of view from which to redefine the task of analysis and psychotherapy. And schizophrenia is 'the absolute decodification of flows'.

Schizophrenia is the exploration of semiotic territories by a non-subjectified flow of assemblages that are not fixed on a definitional refrain. Thus, it does not ossify any model of interpretation and is not identified as a closed subjectivity.

In this sense, while recognizing the suffering of this absolute opening, of this speed of displacement and imagination that does not manage ever to stop and solidify, *Anti-Oedipus* makes of this process its own point of view, from which it observes the agglutination of the normative social field as an effect of a territorialization, of a molar redefinition of identity.

What circulates in the social body, in the process of communication? What is the substance that holds together the symbolic exchange that motivates human actions?

This substance is abstract; it is desire. It does not suffice to say: what moves us is money, or what moves us is religion, or ideology. One must understand how this abstract substance permeates behaviours, how it is transformed into desire.

> The genetic revolution occurred when it was discovered that, strictly speaking, there is no transmission of flows, but a communication of a code or an axiomatic, of a combinative apparatus [*combinatoire*] informing the flows. Such is also the case for the social field: its

> coding or its axiomatic first determine [*sic*] within it a communication of unconsciouses. (*Anti-Oedipus*, 276)

The substance that circulates in the biological body is a semiotic substance that the genetic code comes to define. In the same way, there circulates in the social body a substance able to stabilize human aggregates for long periods of time. Refrains are the DNA of social bodies: they act as organizing devices for the substance that circulates through communication, combinatory styles that permit the recognition of an individual, or a collective, as a singularity.

The concept of refrain is not far from what Gregory Bateson defined as the 'pattern which connects' (*Mind and Nature*, 8–12):

> Interesting phenomena occur when two or more rhythmic patterns are combined, and these phenomena illustrate very aptly the enrichment of information that occurs when one description is combined with another. In the case of rhythmic patterns, the combination of two such patterns will generate a third. (*Mind and Nature*, 79)

The refrain is an assemblage of signs that function in order to give to whoever assembles them the opportunity to come into contact with other men, with the social world, with the cosmos. The refrain is the recognition cipher, a sort of singular code for accessing the cosmos, a vibration that assembles and connects with the surrounding vibrations.

In front of Tinguely's cosmic cogwheels on display at the Pompidou Centre exhibit, Félix stopped for many long minutes in melancholic ecstasy. And then he said that the Tinguely cogwheels were a metaphor for the refrain, devices that try to hook into the cosmos, to enter into communication with the inaccessible.

There is no truth, then, if not singular. And the refrain is the activity of connecting.

But what does all this have to do with '68?

'68 here means the movement of rebellion against all kinds of bondage, not only those of exploitation, of oppression, of imperialism, of the family, but also and especially against the bondage of the Signified. '68 freed up the possibility of building desiring assemblages, of the signifying function of the assemblage:

> If desire is productive, it can be productive only in the real world and can produce only reality. Desire is the set of *passive syntheses* that engineer partial objects, flows, and bodies, and that function as

> units of production. The real is the end product, the result of the passive syntheses of desire as autoproduction of the unconscious. Desire does not lack anything; it does not lack its object. It is, rather, the *subject* that is missing in desire, or desire that lacks a fixed subject; there is no fixed subject unless there is repression. Desire and its object are one and the same thing. (*Anti-Oedipus*, 26, emphasis in the original)

Once it is freed from ideological representations of its protagonists, '68 is the first movement of mass desiring productions, the first refrain of refrains that has ever been felt on the planet: a deafening refrain, a gigantic mechanization of Tinguelyian cogwheels that together conjure up a universe of non-necessary, but possible events. '68 was in this sense the first movement without necessity, without lack, without need.

> [As Marx notes,] what exists in fact is not lack, but passion as a 'natural and sensuous object'. Desire is not bolstered by needs, but rather the contrary; needs are derived from desire: they are counterproducts within the real that desire produces. Lack is a countereffect of desire; it is deposited, distributed, vacuolized within a real that is natural and social. (*Anti-Oedipus*, 27, citation modified to conform to original French)

Creationism

The theme of desire first articulated in *Anti-Oedipus* anticipates the development of subsequent texts. In *Chaosmosis*, Guattari's final work, he speaks of the relationship between desire and rhythm: rhythm is the modality of projecting the world by a singularity. And desire alone can assemble a singular rhythm with another singular rhythm. What Guattari called the refrain is precisely this singularization of rhythm, this singularity of breathing, of striding, of speaking, of gesturing and especially of synchronizing oneself with the world. Desire allows a refrain to tune itself in with another refrain.

When different refrains start singing in tune there occurs a shared world.

We can call this conception creationism because the world appears here like the projection of a creative activity that intersects and assembles with other creative activities. The truth is a bridge over the abyss of sense. But you cannot travel over this bridge if you do not share the intensive world of whoever constructed it in order to traverse the abyss. It is not a question of sharing his reasoning, nor of expressing agreement, because it is not a question of the content of an argumentative activity. It is a matter of conceptual creations, that is, of projections capable of creating a world, of constructing traversable bridges.

The truth does not consist in finding a sense around which to construct consensus, but consists in projecting worlds endowed with sense for whoever shares the intensity that is the motor of this projection.

Guattari felt a certain distaste for the big words of traditional philosophy, such as truth and freedom, and yada yada yada.

******

The idea that the collective process can be configured as the realization of an implicit truth in history is the Hegelian heredity from which the '68 consciousness was unable to free itself. In the period following '68, precisely when it seemed that the dialectical project was about to be realized, the dialectical illusion dissolved and society became ever less reducible to a global project. From that moment onward, speaking about truth – to seek a truth, a finality, a positive destination from the intricacies of collective events – became increasingly improbable. From that moment onward, only the communities of friendship (not of membership) could produce sharable social life.

Unfortunately, in the political history of the last decade, the '68 generation understood this too late. After having hurriedly abandoned the totalitarian idols of the twentieth century, it replaced dialectical truth with cynicism, converted over to any conformism and stooped to any academic, political and spectacular baseness.

But friendship remains the richest lesson of the history of the movements. And friendship means a provisional community that is not based on any common origin, on any written destiny, on any historical necessity, but instead only on provisionally assembling refrains. It means love for the same situations, pursuing the same provisional objectives, taking pleasure in following the same path together, or failing together, and falling.

9
Kafka, Hypertext and Assemblages

> According to Guattari and Deleuze's definition, a 'minor literature' is the literature of a minority that makes use of a major language, a literature which deterritorializes that language and interconnects meanings of the most disparate levels, inseparably mixing and implicating poetic, psychological, social and political issues with each other. In analogy, the Japanese media theorist Toshiya Ueno has recently referred to Felix Guattari as a 'minor philosopher'. Himself a practicing psychoanalyst, Guattari was a foreigner to the Grand Nation of Philosophy, whose natives mostly treat him like an unworthy bastard. And yet he has established a garden of minor flowers, of bastard weeds and rhizomes that are as polluting to contemporary philosophy as Kafka's writing has been to German literature. (Broeckmann, *Minor Media,* online, para 1)

In 1975, Deleuze and Guattari, working within the conceptual context of *Anti-Oedipus*, published a book entitled *Kafka. Toward a Minor Literature*. Kafka is placed under the shadow of Oedipus and comes to be projected into a scenario that goes far beyond the psychoanalytic horizon. 'An Exaggerated Oedipus' is the title of the second chapter, but Oedipus is not the key for interpreting the Kafka text: on the contrary, Oedipus is described as a libidinal machine, an investment of neuroses.

> It is not Oedipus that produces neurosis; it is neurosis – *that is, a desire that is already submissive and searching to communicate its own submission* – that produces Oedipus. Oedipus, the market value of neurosis. In contrast, to augment and expand Oedipus by adding to it and making a paranoid and perverse use of it is already to escape from submission, to lift one's head up, and see passing above the

> shoulders of the father what had really been the question all along: an entire micropolitics of desire, of impasses and escapes, of submissions and rectifications. Opening the impasse, unblocking it. Deterritorializing Oedipus into the world instead of reterritorializing everything in Oedipus and the family. (*Kafka*, 10, emphasis in the original)

Kafka constructs Oedipus as an obsessive cartography, but his desiring investment, his textual libido, is a line of flight that moves in a thousand directions and is not reducible to Oedipus.

The unconscious is not a theatre, but a factory, as Deleuze and Guattari wrote in *Anti-Oedipus*. There are no fixed roles that we would be required to adopt and to perform, but rather parts of a machine, assemblages of enunciation that lock and unlock labyrinths from which one toils to exit, and paths of flight that lead into another territory.

Deleuze and Guattari's *Kafka* is not a psychoanalytical book, nor is it a book about the unconscious in Kafka's literature. I would rather say that it is a book about hypertext. Even if they don't use this word (which was not in use in those days), this is in fact the concept on which they were working.

> The castle has multiple entrances whose rules of usage and whose locations aren't very well known. The hotel in *Amerika* has innumerable main doors and side doors that innumerable guards watch over; it even has entrances and exits without doors ... Only the principle of multiple entrances prevents the introduction of the enemy, the Signifier ... (*Kafka*, 3)

The philosophical imagination of the rhizome is at work here, and the model that it traces is the hypertext, that is, a semiotic construction in which any sign can introduce transversal series of assemblages.

The act of connecting does not assume the form of interpretation, nor is it the explication of something that was already implicitly there. There is no implicit-explicit relation in hypertextual semiotic assemblages, but the crossing of discursive planes, processes of distancing and undistancing, lines of flight, the passage from one plane of consistency to another.

The hypertextual principle is a declension of a-dialectical becoming; the passage from one configuration of the world to the next is not the passage from an implicit state to an explicit one, but a deterritorialization. In the same way, the schizoanalytical path does not consist in the passage from a problem to a solution, but in the displacement of our foci of attention.

Schizoanalysis is the deterritorialization of investments of desire; it is a production of the unconscious capable of modifying the neurotic frame, of opening it so as to include unforeseen objects and determine restructurations of the psychic field.

Schizo-Kafka proceeds through metamorphosis and not metaphor. Gregor Samsa's beetle is not a metaphor, but precisely a metamorphosis.

> Kafka deliberately kills all metaphor, all symbolism all signification, no less than all designation. Metamorphosis is the contrary of metaphor. There is no longer any proper sense or figurative sense, but only a distribution of states that is part of the range of the word. The thing and other things are no longer anything but intensities overrun by deterritorialized sound or words that are following their line of escape. It is no longer a question of a resemblance between the comportment of an animal and that of a man; it is even less a question of a simple wordplay. There is no longer man or animal, since each deterritorializes the other, in a conjunction of flux, in a continuum of reversible intensities. (*Kafka*, 22)

Following the path traced by Kafka, Deleuze and Guattari come to the elaboration of their own poetics of the minor. Minor literature, they write, is not the literature of a minor language, but the literature that a minority makes in a major language.

Against any populism or folklorism of resistance, they state here that the sole terrain in which sabotage works effectively is the most advanced, the most complex one. Enunciation must assemble elements lifted out of the enemy's territory because, above all, minor literature acts as 'a high coefficient of deterritorialization' (*Kafka*, 16–18).

> In this sense, Kafka marks the impasse that bars access to writing for the Jews of Prague and turns their literature into something impossible – the impossibility of not writing, the impossibility of writing in German, the impossibility of writing otherwise. (*Kafka*, 16)

The language in which the minor literature locates itself is the major language, and only out of its tangle is it possible to construct rhizomes that move outside the language, outside the order that the language protects; only out of the chaos of a language (that still pretends to be ordered, grammatical, structured), only out of this chaos in language does it become possible to create refrains able to assemble, to put in motion collective deterritorializations, and to open the universe of singular sense.

Are we not focusing perhaps on the same problematic, the literary blossoming that in the last decade was called (inadequately) *postcolonial*? Kafka wrote in German as Naipaul wrote in English, not because those were his roots, but precisely because it was his rhizome. Rushdie, Kureishi, Bharati Mukherjee also write in a major language to construct their labyrinths of glances. Minor literature does not seek to constitute an identity, nor does it try to rediscover it; it does not seek to reterritorialize signs through the imposition of the Signified. There is no identitarian intention in minor literature, but a path of singularization starting from a post-identitarian condition.

Minor literature is the gaze from the outside, the gaze of somebody who observes the ritual without knowing the code and thus understands its a-signifying nature.

> How many styles or genres or literary movements, even very small ones, have only one single dream: to assume a major function in language, to offer themselves as a sort of state language ... Create the opposite dream: know how to create a becoming-minor. (*Kafka*, 27)

Becoming-minor – today, at this end of the century in which aggressive identities are being consolidated, anxious to create a territory in which to impose violence and authority – is the sole tolerant hope, the sole path leading to a happy drift.

In *postcolonial* literature, two directions confusedly overlap: one is that of identitarian claims, the rancorous reaffirmation of a community of roots. The other is that of happy non-identity, of the provisional community of desire. The collectivity of enunciation is a fragile architecture based on the sharing of a refrain, on the creation of a path of sense not guaranteed by any root, by any integrity, by any violence, but only by friendship and its provisional eternity.

The possibility of assemblage is based on the provisional eternity of friendship.

What is an assemblage (*agencement*)? It has nothing to do with contradiction or dialectical connection, nor with an implication of homogeneity or opposition.

> On the one hand, [the assemblage] is segmental, extending itself over several contiguous segments or dividing into segments that become assemblages in turn ... The segments are simultaneously powers and territories – they capture desire by territorializing it, fixing it in place, photographing it, pinning it up as a picture, or dressing it in tight

> clothes, giving it a mission ... But [on the other hand] we must declare as well that an assemblage has points of deterritorialization; or that it always has a line of escape by which it escapes itself and makes its enunciations or its expressions take flight and disarticulate, no less than its contents that deform or metamorphose ... (*Kafka* , 85–6)

The assemblage is not a process of interpretation or the discovery of something waiting to be discovered, a synthesis of opposites or the realization of a putative truth. The assemblage is not a theatre, but a factory. It acts on segments, extensive and intensive material elements that have no homogeneity. Place together electrons, silicon, telephone lines, human brains, and from these you can make a digital web.

Not necessarily. Not because the digital web is the logical deployment of the elements that constitute it, but because machinic assemblages that ignore roots only know combinations.

This is why the principle of the assemblage puts into motion an unlimited creativity.

> The assemblage extends over or penetrates *an unlimited field of immanence* that makes the segments melt and liberates desire from all its concretizations and abstractions. (*Kafka*, 86, emphasis in the original)

The emergence of harmony from chaos (chaosmosis) results from the creativity of assemblages, from outlining constellations in order to recognize in this the sole possible bridge over the abyss of senselessness.

Tantric egg

The reading of *Anti-Oedipus* has already raised the idea that the machinic and the organic are not two distinct domains, two incompatible systems, but rather provisional molar fixations of molecular and submolecular material matter. The machinic, the organic, and the semiotic combine together in the unconscious, and the rule of combination is not definable in the abstract as a universal generative grammar. The unconscious arranges organic, historical and sexual segments.

The distinction between machine and organism, between life and mechanical assemblage, is only a way of speaking, a discursive and classificatory strategy, an epistemic expedient that we must constantly put into question on the sub-segmental and sub-atomic levels.

> Jacques Monod has defined the originality of these syntheses, from the standpoint of a molecular biology or of a 'microscopic cybernet-

> ics' without regard to the traditional opposition between mechanism and vitalism. (*Anti-Oedipus*, 288)

The traditional opposition between mechanicism and vitalism reveals itself as too limited to comprehend the development of microphysics. On the sub-atomic scale, the distinction between domains of matter is losing its meaning.

Modern humanism fixes a rigorous hierarchy of different domains of matter because this is functional for its epistemic strategy. The domain of inorganic matter had to be distinguished from the domain of organic matter, and then a special place was given to the human domain. Today we are currently oriented toward a different epistemic strategy in which the human no longer maintains any privileged place, and especially can no longer appeal to a sort of hierarchical separation.

We hear people talking about the post-human, and this expression can mean many things.

Above all, it can mean the failure of rationalist humanism that has been unable to govern the social relations between organisms of the human kind, and today it is constrained to surrender in the face of the unleashing of the bestial, the pulsional and the machinic.

One can speak of post-human also in the artistic sphere because artists, who arrive more quickly and courageously than the politicians at the inevitable but embarrassing consequences of historical failures, have begun to stage the human body as a carnal segment in a heterogeneous machinic continuum in which organic matter, inorganic matter and digital pulsations interact in the form of a bio-complex organism. In recent years, bioengineering has generated doubts about whether the human marks an absolute discontinuity in nature.

It is precisely because it problematized the material continuity between the organic and the machinic that *Anti-Oedipus* anticipated post-human thinking by two decades.

But it is especially *A Thousand Plateaus*, eight years after *Anti-Oedipus*, that muddies the waters of any classificatory distinction in the material sphere. Matter, in *A Thousand Plateaus*, is a flow in continuous recompositional becoming.

A treatise on semiochemistry[1]

We can consider *A Thousand Plateaus* to be a treatise on semiochemistry even if, as far as I know, Deleuze and Guattari never used this term.

Signs are not (only) signifying, but also material segments, factors in deterritorialization, viral agents, strategies of displacement.

With the expression semiochemistry, it is possible to synthesize the sense of the entire discourse on the double articulation of language developed in *A Thousand Plateaus*. In language we can distinguish a plane of the sign and a plane of the event. The words, the sound, the grapheme, the image, are all signs and they all open up signifying chains. But events are also material things, objects that function according to particular laws and produce specific effects.

This field was sown by Russian formalism at the start of the century, and the artistic avant-garde reaped its fruits throughout the twentieth century. If we take note of this double articulation in language, we can distinguish a semantic level – a level on which signs produce signifying effects – and a semiochemical level on which signs function affectively, opening emotional, relational, imaginary chains. Signs themselves behave as if they were events, and events behave as if they were signs.

In an essay in the volume, *Il secolo deleuziano* [*The Deleuzian Century*], entitled 'How Deleuze Makes Signs', Paolo Fabbri says that:

> Any sign is the effect of the action of a body on another body, and therefore *affect*: and this variation of effects on a body provokes a variation in power [*potenza*], in affective sensibility: increase of power (joy), decrease of power (sadness) ... Always relying on Spinoza, Deleuze asks the question of the traditional distinction between conventional sign and natural sign, that is, the problem of the arbitrary aspect in language. And Deleuze's answer is obvious: no, language is not arbitrary given the obvious consideration that linguistic signs always are connected with other signs that are natural signs ... Any sign is a translation of other signs, and above all, signs always refer to signs. In short, there is no external reality to signs since signs are constitutive of objects and nominatives of events, and thus are themselves their own reality. (113–14, emphasis in the original, translation corrected to conform to original)

In the interview 'On Philosophy' (*Negotiations*, 135–55), Deleuze said that the sign can be seen as an incorporeal carrier of meanings. But it can also be considered as a corporeal agent, like a virus, to use a metaphor that William Burroughs used for the first time (see, for example, Burroughs, 'Ah Pook Is Here').[2] There is an acoustic materiality, a graphic materiality, and even a performative materiality of words. Letting out a cry in a crowd is like handling a material substance, it's like handling explosives. And contamination between art and advertising in the media era has shown how signs can unleash gigantic economic processes.

Words are viral agents, as are images and sounds. This does not exclude the possibility that they 'mean something', that they remain within a signifying sphere. When we look at them insofar as they have meaning, they are transparent. This sign interests us because it points to a referential sphere. But at another moment we can consider the sign as a replicant, a mutagenic agent, an event that is assembled with other events. In this case, we cannot seal off separately the sphere of words from the sphere of things because words act as things through other things, place processes into motion and create communication. They are not limited to signifying; they communicate.

As viral agents, they produce mutations. Semiochemistry is the process through which signs produce effects of decomposition and recomposition in the social psyche, in the imaginary, in the wait for different worlds, in desire.

This double articulation allows us to understand also how thought functions, and the thought of Deleuze-Guattari in particular. It functions, of course, as abstraction and interpretation of symbols through other symbols. But at a certain point, the interpretative machine leaves the field to neologisms and contaminations, and the words of philosophy become pop discourse. Alongside argumentation, another kind of functioning is revealed, one that is much more material, dynamic and teeming with life.

There is an immediate, spontaneous proximity between the thought of Deleuze and Guattari and real movement, the becoming-sensible of society. Their books have recorded the subtlest and most intricate displacements of social affectivity in the late-modern period not through analyses of processes, but through the invention of concepts and terms which were able to illuminate processes from innovative perspectives.

In the 1970s, *Anti-Oedipus* recorded the proliferation of micropolitical movements and the formation of a social field of creative work.

In the 1980s, *A Thousand Plateaus* anticipated the proliferation of processes of subjectivation, the decadence of the centralized and linear model of communication. In a certain sense, *A Thousand Plateaus* is the most lucid anticipation of what would take form in the 1990s as Web culture.

The homology between Web culture and rhizomatic thought is not merely a superficial phenomenon, but it concerns the modalities of discursive proliferation through the Web themselves. Mailing lists, discussion groups and hypertextual Web interactivity are nothing other than a teeming projection of the concept of rhizome.

If a connection exists between the thought of Deleuze and Guattari and the becoming-sensible of society, we must not consider it an extrinsic fact, an accidental political-activist adaptability of their concepts. This thinking functions like a semiochemistry, and that is why it is close to real movement. This is a thought that doesn't conceive itself as a coherent system, but rather as a mutagenic apparatus, one that acts linguistically, conceptually, neologistically on the collective conscious organism with which it enters into contact. This is a thought that functions by unleashing an unpredictable chain of links.

Molecules and events

'DG thought' is a compound combination of molecularism and eventualism, that is, a semiochemical viral apparatus.

We can try to see how Gilles Deleuze and Félix Guattari produced both together and separately. What did Deleuze bring to rhizomatic thought with his work on the history of philosophy and particularly on Spinoza, Nietzsche and Bergson? And what did Guattari bring with his work on institutional psychoanalysis, on collectivities of enunciation, on political subjectivation, and on molecular desire?

Deleuze brought the event, Guattari brought the molecules.

Deleuze brought a deconstruction of ontology: the relation between Being and Subject that is the basis of metaphysical construction from Plato to Hegel is dismantled by Deleuze piece by piece in his pre-1968 books, in his studies of Hume, Nietzsche, Bergson, Spinoza and Kant.

But it is Guattari who brought a whole language deriving from chemistry, alchemy, pharmacology, hydraulics and even electricity, as well as, obviously, from political activism.

And it is thanks to all of these conceptual implements contributed by Guattari that the rhizomatic machine begins to operate, because it is through this whole machinery of flows and swarms, it is thanks to this biological and psychochemical teeming, that one can imagine processes of subjectivation, where the history of philosophy had only seen the Subject, and a multiplicity of reversible and non-totalizable becomings (*les devenirs*) where the history of philosophy had situated Being.

******

Deleuze and Guattari's radical innovation starts with *Anti-Oedipus*, where the work that each of them had done separately starts to bear fruit on a ground that no longer separates philosophy and other mechanisms, but allows philosophy to function precisely as a mechanism.[3]

Anti-Oedipus brings to conclusion the path that Deleuze had developed while reading Nietzsche. In *Nietzsche and Philosophy* (1962), Deleuze began to transfer thought from interpretation to production, from work to pleasure, from responsibility to dance. But to complete this trajectory of de-responsibilization, what was needed was the new notion of the unconscious that Guattari had elaborated in his process of taking distance from Freud. So then, from the theatre of traditional philosophy and Freudian psychoanalysis, the factory of *Anti-Oedipus* emerged, a laboratory in which thought starts functioning as a schizo Unconscious that does not pretend to pursue a pre-established programme, but rather to dis-connect (*de-lira*), that is, to transfer its reading from one semiotic plane to another.

Anti-Oedipus brings to fruition Deleuze's work on Nietzsche and Guattari's work on collectivities of enunciation.

What did Deleuze do without Guattari? He meandered nimbly through philosophers with the intention of never contradicting them.

What is the point of contradicting someone's thought, Deleuze asked? Contradiction serves no purpose. This is not an issue of politeness, but the simple fact that Deleuze wandered through philosophers in search of elements that could function in a process of heterogeneities that will be machinically assembled. This is the philosophical creation of concepts and the becoming-event of signs. This is why Deleuze wasted no time on contradiction because he was interested in understanding from the works of different authors what could be reinvented within the semiochemical process. And in the course of this process Guattari appeared.

Deleuze blazed his own trail towards an interaction with Guattarian molecular thought in which biological and specifically pharmacological concepts proliferate virally.

The encounter between Deleuze and Guattari occurred in 1968 when, besides so many other interesting things that occurred, Deleuze published his book *Bergsonism*.[4]

It is precisely here, in the way that Deleuze reads Bergson and that Guattari conceptualizes enunciation, as a mutagenic machinism, as a viral pragmatism, that we can find the philosophical junction between the two rhizomaticians.

Guattari's convergent pathway comes from militant activity and analytical practice. On this level, Guattari applied a cognitive method that is no longer dialectical, but essentially molecular.

Guattari addressed the decomposition of the subject, a mutation in the very concept of the subject. It is not the molar constitution of the subject that interested him, Guattari said, nor the Freudian universal Unconscious mathematized by Lacan, nor the molar dialectic of classes. What interested him was the process of subjectivation that always follows singular pathways.

There is no subject, but rather the process of subjectivation. Subjects become through the juxtaposition of psychic, linguistic, imaginary, biological and historical elements. And this juxtaposition is molecular in character, proceeding through the encounter and conjugation of heterogeneous components (biological, mediatic, logical, affective, machinic and organic). Guattari abandons the universalism of the dialectic and of psychoanalysis by opposing the idea of a constituted subject with the idea of a process of subjectivation having a molecular, singular and combinatory character.

******

In *A Thousand Plateaus*, the Guattarian molecular inspiration meets the Deleuzian principle of plural ontology. In the concept of the Body without Organs (at the centre of *A Thousand Plateaus*), there is a Deleuzian inspiration, the idea of a projective and multiple ontology, and there is also a Guattarian inspiration in the idea of molecularity, of the subsegmental proliferation of matter.

Up to *Anti-Oedipus*, Guattarian thought is a work of decomposition of the social, the psychic and the linguistic object. Guattari worked on this compositional (or decompositional) cognitive model of language and of sociality, seeking to find (through his text, *Psychanalyse et Transversalité/ Psychoanalysis and Transversality*) which psychic segments are at work in the formation of subject-groups and subjugated groups, which social segments are at work within the formation of the Unconscious, and so forth. Social processes, just like natural processes, cannot be explained on the basis of identitarian molar oppositions, but on the basis of the proliferation of mutagenic agents, of their decomposition and recomposition.

The Body without Organs

Let us take this most mysterious and strangest concept, the Body without Organs, that Deleuze and Guattari adopted from Artaud in order to create a kind of metaphor for the mind-body-cosmos relationship.

> We treat the BwO as the full egg before the extension of the organism and the organization of the organs, before the formation of the strata; as the intense egg defined by axes and vectors, gradients and thresholds, by dynamic tendencies involving energy transformation and kinematic movements involving group displacement, by migrations: all independent of *accessory forms* because the organs appear and function here only as pure intensities. The organ changes when it crosses a threshold, when it changes gradients ... The tantric egg. (Deleuze and Guattari, *A Thousand Plateaus*, 153, emphasis in original, translation corrected to conform to original)

The concluding formula, tantric egg, sounds mysterious and suggestive, and recalls the metamorphic wisdom of tantrism, the action that consciousness can exercise on the body.

Both Gilles and Félix always had a certain intolerance for mystical inclinations, but it would be interesting to study the terminological links and also the philosophical implications of their thought with Buddhism.[5]

In the Body without Organs concept, the consonance is obvious: the BwO is the immanence where pure intensities and non-codified and unrepeatable singularities flow. These are impermanent singularities: in its various versions and its very diverse schools, Buddhism is never far from an essential idea, that of the impermanence of all worlds of existence and projections. And the other side of impermanence is the foundational truth of the void.

The visible and experiential forms are only hallucinations of the mind. And even the ego, to which Western metaphysics devotes a kind of philosophical adoration, seems for Buddhist thought only to be an effect of language, the projection of a linguistic and perceptive illusion. The world – that the pre-Buddhist Hindu tradition defined as *maia*, that is, illusory projection – is understood by Buddhism as a succession of mental variations.

In the definition of the Body without Organs as distributor of intensities that trace the lines of the cosmic world, there is a reflection of the Buddhist vision. It is at the junction between the rhizomatic reading and the Buddhist reading that I came to understand the notion of the Body without Organs as an egg, as the surface of an egg on which we find in the state of pure potentiality the lines of flight or lines of formation that will contribute to the indeterminacy of subjectivation. The word 'subject' in Deleuze and Guattari is always viewed a bit suspiciously. There is never any subjectivity, but always subjectivation. Any instance of subjectivation is a Body without Organs, that is, an unexpressed potentiality and becoming.

Subjectivation presupposes the subject in an illusory way, but in reality, it places into being the subject's provisional determinations even as it immediately surpasses and recombines them. This is also how the Body without Organs presupposes the body but, at the same time, determines its articulations in physical space through intentionality and relations of semantic proximity with the other.

******

Even here I see a question of semiochemistry: subjectivation in fact is the assemblage of the enunciator through the process of enunciation. We think that we speak through the will of our subjectivity, but in reality, the language that we are speaking speaks to us, and the relations in which we are caught are our process of subjectivation. There is no subject of enunciation that exists before the enunciation itself. It is the enunciation that brings into being the organism as subject, that subjectivizes us; and any subjectivation is the start of a new process of dissipation bringing us back to the condition of egg, of Body without Organs, of becoming other.

The expression 'becoming other' always refers to the same problem: subjectivation is becoming other, that is, the interaction between indeterminacies that are provisionally determined precisely through their interaction. In this way we can define the Body without Organs.

The Sanskrit word *asamskrta* itself defines this very thing, the indeterminacy, what has not attained the state of determinacy.

Whatever has not attained the state of determinacy is one among innumerable states of becoming other.

A subject of enunciation does not exist; what exists is a reciprocal implication of the enunciation and of the process that enunciates.

The Body without Organs is so difficult to define simply because it concerns the very undefinability of the subject. And when I say undefinability, I am not referring to a vague and undetermined condition, but rather to a condition of possibility, exactly to the egg. Because on the egg we can already read a map of intensity, of directions, of lines of flight. This is the egg: not yet a subject, not yet a constituted process. Neither is it a code that predetermines the organism's fate, but only a limitation of the infinite field of possibilities.

We can here glimpse the whole question of the relations between genetic and epigenetic codes, between the molecular structure of DNA and the development of the organism.

Language and life

Biogenetic research shows how the formation of the organism (the essential process of biology) can be described as a linguistic process, as the encoding of matter by a kind of linguistic program.

A new paradigm of a bioinformational sort has seemed to emerge from the convergence of information technology and biotechnology. In the first place, this happens because the computer is indispensable for reading and archiving the molecular sequences that compose the organism's programme. But there is also a deeper analogy between language and life, between phenomena of communication and of epigenesis.

The matter from which organisms are made seems to be penetrable by the substance of signs (or better, by signs understood as substance). DNA possesses this double characteristic, of being at the same time substance, organic matter and sign. It is a programme that replicates itself in the organism's matter, that governs processes of interpretation, and that can be read by a computer only because the organism can read it too. Language presents itself here in its double articulation, as representation but also as matrix.

> A semantic level, as in linguistics, appears in the protein's signification in the cell, specifically its function (which is not directly visible in the sequence of amino acids, like the signification of a word is not visible in its writing or in its pronunciation). The rules for rewriting the genetic patrimony from DNA to the proteins are summarized by the correspondence of the *genetic code*. (Antoine Danchin, *La Barque de Delphes*, 29, our translation, emphasis in the original)[6]
>
> It is the *correspondence* between the sequences of nucleotides and those of the amino acids, which changes the operating level of the physical entities from that of the nucleic acids to that of the proteins, that makes it possible to use the metaphor of a program when we talk of the *formal* content of DNA. The physical creation of an individual organism is determined by the consequences of the expression of its genetic program, carried out by those 'manipulating' objects the proteins (which are in particular capable of manipulating the very

> substance of the program, the DNA, by introducing variations, while copying it) (Antoine Danchin, *The Delphic Boat*, 256, emphasis in the original).

Epigenesis (the process through which the organism is generated through successive passages of cellular replication and specification) can be described as a process of executing a programme inscribed in the genetic text. But as much as an information technology based on epigenesic reading might be possible, we must not forget that in this case information is specified and developed through biotic matter (proteins, the environment, material causality).

This means that the information contained in the genetic code is not enough to explain life because the organism's development occurs in environmental conditions that singularize it.

Information, as the algorithm of infinite possible processes, is a pure relationship while life is a concrete becoming; it is the specification of information through the causal materiality of unpredictable processes. Information does not die. It is a pure relation while life tends toward death as the dispersion of information in time.

The event is the material decaying of information. The DNA does not suffice to create *bios*.

Information is but a part of the process of the organism's specification, and the alphabetical metaphor of heredity thus reveals itself to be something more than a simple metaphor.

Embryonic cells can be compared to a row of computers that operate in parallel and exchange information with each other. Each cell contains the same genome and thus the same written program. But it can verify a quantity of different states, and the program directs its development along different alternative paths, depending on the combination of flowing information that the cell has in its memory and following the current environmental signals that it receives.

On a certain semiotic and biotic plane, they cannot function without each other. And this plane is what Deleuze and Guattari call the Body without Organs.

How can writing create a world? Biologists are asking this today, as do all those who are studying the relation between DNA information sequences and the development of the cell in an organism.

> The role of the coding process is to make the transfer from a chemical world in which, broadly speaking, the objects (in this case segments of DNA) can be regarded as exploring only one dimension of space,

> to a world in which other objects, proteins, explore it in three dimensions, or even four if we include time, because proteins can change their shape. (Antoine Danchin, *The Delphic Boat*, 284)

Danchin compares epigenesis to the generative power of writing.

Writing does not produce a world as an effect of information, but through intersections of waves. Try to picture how a wave is produced in a sports arena, as Alessandro Sarti suggests.[7] Someone raises his hands, others beside him do the same thing, people on the right lift their arms while the first ones drop them, and then the wave spreads out toward the right, and row after row a hundred thousand spectators raise their arms in slow or rapid succession, and then more slowly or more quickly.

How can a similar phenomenon happen (this is what Kevin Kelly also studied in his *Out of Control*)?[8] Between information and matter, between codes and the event there is a relation that resembles the form of the wave. What is a wave? It is an accumulation of modifications of the status of a fluid body, transfers of form, information that moves and is moved.

The wave consists of an intersection of information and fluid matter.

Even literature produces waves in the sense in which Burroughs said that Jack Kerouac caused a million blue jeans to be sold and that Mayakovsky moved proletarian troops.

This does not happen because words create consensus and action, but because words accumulate in communicative space until they create an anticipated rhythm, a way of being, desire.

Here we enter into the problem of the semiotic double articulation, that of the dual sphere in which linguistic signs function, that is not only the semiotic sphere, because words are also psychic stimuli, graphic matter that ends up modifying the ethno-mediascape.

> A semiotic chain is like a tuber agglomerating very diverse acts, not only linguistic, but also perceptive, mimetic, gestural, and cognitive: there is no language in itself, nor are there any linguistic universals, only a throng of dialects, patois, slangs, and specialized languages. (*A Thousand Plateaus*, 7)

Assemblages of a-signifying traits

The distinction between the biological, the semiotic and the machinic has an intellectual and cognitive usefulness, but in reality, distinct and hierarchically separate spheres do not exist. When we move in the world of affective phenomena, the biological, machinic and semiotic implicate and criss-cross each other. Heterogeneous fragments compose each other on singular planes of consistency.

The distinction between the biosphere and the noosphere is useful at a very abstract level, but does not account for the fact that biological evolution incorporates mutagenic elements coming from the noosphere as well as machinic elements coming from the technosphere.

The viral principle is irreducible to mono-planarity since cultural viruses can be transferred into the biological universe on the one hand and on the other hand biological viruses can produce cultural effects, as demonstrated by the effects that the HIV virus has had on desire, the imaginary, culture and daily life. There is no biosphere, noosphere or any other sphere that is distinct and circumscribed, but rather a kind of mechanosphere.

The mechanosphere is the sphere in which entirely heterogeneous elements are assembled and still capable of functioning together. We can call these elements different things: Guattari himself used the expression 'a-signifying machinic traits', meaning elements of heterogeneous types (material, imaginary, biological, noetic ...) that in themselves cannot create sense, but conjoin to constitute coherent constellations that can define events, or haecceities: singularities endowed with meaning.

To understand what a-signifying machinic traits might be, we can think about how viruses function. A virus does not manifest its behaviour (its possibilities, its meanings) without being in relation with an organism; then we have effects of mutation, illness, morphogenesis.

We say a-signifying traits because these affective, linguistic, technical, biological, ideological, imaginary, etc., segments do not carry along with them their own fate or finality, and do not go in one direction rather than another. They acquire their meaning in mutable relations with other elements. Traits of this kind can function according to contradictory modalities.

There are behaviours, symbols and enunciations that can function as consolidating elements of the fascistic obsession with identity, or can free up schizoid processes and even put into motion liberating processes, depending on the assemblages in which they are caught and on the aggregate they end up composing. Any a-signifying machinic trait produces different effects depending on the inter-machinic complex with which it reacts, and on the assemblage in which it gets caught. For example, the feeling of belonging within a community can produce an effect of solidarity and friendship in certain assemblages, but in others can produce effects of aggression against the external community. This does not only depend on political, community-related or affective factors, but also on all sorts of heterogeneous elements such as mythological sedimentations, technical equipment, the communicative modalities of the community, and so forth.

When they define the rhizome, Deleuze and Guattari describe it in this way:

> The principal characteristics of a rhizome: unlike trees or their roots, the rhizome connects any point to any other point, and its traits are not necessarily linked to traits of the same nature; it brings into play very different regimes of signs, and even nonsign states. (*A Thousand Plateaus*, 21)

Social evolution depends precisely on assemblages of heterogeneous elements that we believe to be able to identify by defining a stratum of being, but that instead are sucked in by the stratographic superposition that is becoming. As theory and practice, politics simplify and crystallize the complexity of the process. Political reason believes that what determines the historical process are strategies of intention, structuring forms of will, militant or military deployments, states, apparatuses, conscious choices, consensus. And yet, based on these things, one cannot explain many significant events, the rupture from which new series of phenomena are born or the movements that upset the consolidated institutional structures. In reality, strategies (and particularly, political strategies) function as adaptive reactions to automatisms in the technical realm, in finance, in techno-communication, in mythology, and in the social psyche.

Social changes do not depend on the will to change or on the clash of different wills, but on the insertion of elements that were not included, or even foreseen, within the process. An imaginary obsession, a refrain that is diffused, a new technological mechanism, arriving from a marginal or foreign stratum with respect to the political sphere, can all completely modify the existing social assemblage, following lines on which the will can no longer act except marginally as each of the a-signifying traits enter into complex processes. Who would have thought that a marginal practice like digital networking would finally transform the global economy? Who would have thought that at the end of the modern era, archaic religions would become the decisive factor in the international political system?

*****

Deleuze and Guattari do not establish any hierarchy or any evolutionary direction that would bring us, for instance, from matter to mind or from multiplicity to unity, nor do they create taxonomies capable of imposing an inalterable order of things and concepts. They write: 'The plane of consistency knows nothing of difference in level ...' (*A Thousand Plateaus*, 69).

While a difference of level exists between the biological and the noological spheres, for example, there can always be a plane of consistency that selects segments with biological valence or segments with cultural, spiritual, or ideological valence. Then we could define the plane of consistency as a particular, singular slice that operates through a constellation of events, signs, or affects. Any essentialism, then, gets definitively dissolved because beings are no longer definable as belonging to an archetypical series; there is no reference to the essence of the event or of the aggregate. Events, aggregates and signs do not belong to any series that reduces them to any sort of archetype. They are to be considered rather as singular constellations, singularities composed by traits that have been borrowed, copied or even ripped out from other singular constellations.

Like in the Democritean conception, phenomena are considered as aggregates of entirely provisional elements.

With the expression 'molecular revolution', Guattari synthesized his way of thinking about politics and social becoming. Modes of social behaviour, imaginary stratifications, desires and obsessions conjoin molecularly and not in identitarian form. There is no constituted subjectivity bearing positive values that would be opposed to a social subject defending negative values. There is no dialectical contraposition, but the continuous dislocation of imaginary fragments that can be put together either according to identitarian, aggressive rules, or to desiring, dispersive and politically libertarian rules.

The compositional elements (let's call them pure atoms, molecules, or better yet, viruses, contagious signs as messengers of information) are not significant in themselves. They acquire semiotic power in their composition with other elements:

> Lines of signifying decoding, composed of discrete figures – binarizable, syntagmatizable, and paradigmatizable – sometimes appear in one Universe or another. And we can have the illusion that the same signifying network occupies all these domains. It is, however, totally different when we consider the actual texture of these Universes of reference. They are always marked with the stamp of singularity. (Guattari, *Chaosmosis*, 38)

What we call reality is therefore the projection (or rather the innumerable successive projections) of a functional or semantic continuity on the constant becoming of constellations, on the incessant deterritorializations of a-signifying segments.

In a talk at a meeting on 'Psychoanalysis and Semiotics' held in Milan in 1974, Guattari spoke about signifying semiologies and a-signifying semiotics:

> My opposition between despotic signifying semiologies and a-signifying semiotics remains highly schematic. In reality, there are only mixed semiotics which belong to both in varying proportions. A signifying semiology is always shadowed by a sign machine, and, conversely, an a-signifying sign machine is always in the process of being taken over by a signifying semiology. But it is certainly useful to identify the polarities represented by the two, in other words the signifying semiology as a paranoid-fascist ideal, and a non-signifying semiotics as the ideal of schizo-revolutionary diagrammatization, of getting beyond the system of signs towards the plane of consistency of particle-signs. ('Subjectless Action' in Guattari, *Molecular Revolution,* 140)

Guattari uses here the notion of the particle-sign. This is his way of molecularizing semiosis, of seeing it as an activity of projecting psychochemical agents. We can speak of signs as material agents of semiotic mutations. This has nothing to do with ethereal messengers of meaning, but instead with viral agents in Burroughs's sense who, in fact, spoke of language as a virus: signs not as pure representations, but propagating as a signifying contagion, transforming the semiotic ecosphere.

*****

What does planetary becoming have to do with the Body without Organs?

There is no planetary becoming, Deleuze and Guattari would reply. There are only 'becomings'. At the start of *Chaosmosis*, Félix Guattari writes that in the miasmas and nebulae at the end of the twentieth century, in the middle of the frightening progress of processes of fascistic reterritorializations that seem to dominate our era, we can find a line of flight residing uniquely in the possibility of a creationist chaosmos that is emerging from a multifaceted planetary becoming.

Creationist chaosmos means precisely this: a planetary becoming does not exist; there is no consistency between one state of the world and another because in any process in which we enter, what is interesting is revealed precisely as being inconsistent, it is that which is not consistent and thus cannot be presently foreseen, nor can it ever be analysed as a necessary implication of the process.

If we must base our hope on coherent possibilities within the actual state of the world, if we must expect an alternative at present logically and historically implicit in the state of the present world, we will have nothing at all left but despair. There is no predictable salvation in the present state of the world. Only the unpredictable, what is yet to happen, what is yet to be written into the present plot of the known can inaugurate an entirely inconsistent series with respect to the catastrophe of late modernity.

The surprise consists in the fact that there is no consistency in the historicist sense between a state of the world and a successive one, but there is the tantric egg: the uninterrupted generation of unpredictable constellations, the chaosmotic creation that at every instant invents the world from nothing.

10
The Tantric Egg

In *Anti-Oedipus* we read:

> The body without organs is an egg: it is crisscrossed with axes and thresholds, with latitudes and longitudes and geodesic lines, traversed by *gradients* marking the transitions and becomings, the destinations of the subject developing along these particular vectors. Nothing here is representative; rather, it is all life and lived experience: the actual, lived emotion of having breasts does not resemble breasts, it does not represent them, any more than a predestined zone in the egg resembles the organ that it is going to be stimulated to produce within itself. Nothing but bands of intensity, potentials, thresholds, and gradients. A harrowing, emotionally overwhelming experience, which brings the schizo as close as possible to matter, to a burning, living center of matter. (*Anti-Oedipus*, 19, emphasis in the original)

The Body without Organs is the condition of desiring possibility, a confusional state of matter in which later possible states are contained in an undifferentiated manner. But any state of matter, any state of being contains the later states as actualizations of a virtuality, as a becoming-organism of the egg.

To explain the meaning of the Artaudian notion of the Body without Organs, Félix and Gilles turn to a comparison with emotion. The emotion provoked by an object, an event, a situation, does not resemble this object, this event, or this situation. The emotion provoked by the vision of a breast, by the desire to touch it, to caress it, to suck it – all of these emotions do not resemble the breast, are not facts of its very matter, if – here is the Body without Organs – we do not refer to the pre-actualized, pre-specified matter which constitutes desire, that is, a becoming-other.

And in *A Thousand Plateaus*, we instead read:

> We treat the BwO as the full egg before the extension of the organism and the organization of the organs, before the formation of the strata; as the intense egg defined by axes and vectors, gradients and thresholds, by dynamic tendencies involving energy transformation and kinematic movements involving group displacement, by migrations: all independent of *accessory forms* because the organs appear and function here only as pure intensities ... The tantric egg. (*A Thousand Plateaus*, 153, emphasis in the original)

We can consider the tantric egg as the rhythm that allows a syntony between *prana* and *atman*, cosmic breath and singular soul. It is the locus of indetermination, the condition of becoming-other. Consciousness is constituted by the same elements of which rocks, butterflies and integrated circuits are made, because in each of these worldly objects the same undetermined potentiality circulates, the Spinozist substance.

But this is never tinged with any form of spiritualism because phenomena of contiguity and osmosis between one state of being and any other state of being are made possible by matter itself, by the dynamics that animate it, the manner in which it gives life to its more or less stable configurations, which in the last analysis are all unstable (from the point of view of a Time without Organs).

******

Ah, I was forgetting – sooner or later we will have to talk about the Time without Organs, that is the independent time of duration, the time that is not incarnated in the pulsations of the sentient organism, but is the condition of existence of the body and of its pulsations. The time of the eternal return.

The biospheric-noospheric continuum cannot be defined either as *bios* or as *noos* because the inorganic proliferates in the organic, the organic proliferates in the noetic, and the noetic in its turn circulates prolifically in the organic (as communication psychochemistry teaches us), and the organic in the inorganic.

The telematic web is woven from organic brains, human thought, wires and electrons, ether and silicon. It's the flow that connects all of this.

We can speak of this interchange in terms of a general semiology as if the world were the locus of signs, but we can also speak of it in terms of a generalized biology that penetrates machines, connects the intelligent and the inorganic, transforming connectivity into a self-replicating mechanism of a bioartificial order.

But we can also speak of it in terms of a general geology because in the organic is the inorganic and in the noetic is the inorganic-organic.

But above all, we can speak of it in terms of Mechanology, that is, the science of assemblages.

What is a machine? A machine, Guattari answers, is an aggregate that functions independently of the consistency and the meaning of the elements that contribute to its constitution. A machine is the semiotic functioning of a-signifying fragments that acquire their signifying valence from the fact of entering into connection.

On the one hand, desire is the machinic creation of the elements that remain a-signifying as long as they are fragmentary, distinct, and on the other hand, we cannot consider the concept of the Body without Organs defined in a truly conclusive way.

> You never reach the Body without Organs, you can't reach it, you are forever attaining it, it is a limit. (*A Thousand Plateaus*, 150)

It is useless to tire oneself too much by attempting to find a definition of the concept of the Body without Organs. It is an indefinable notion precisely because it concerns the theme of indefinability, of indeterminacy, and of dis-identity.

The Body without Organs is the condition of possibility of becoming other; it is the presence of everything in everything that seems to be a consequence of the eternal transmigration of heterogeneous particles from one organism in composition-decomposition to another.

This is why I find that there is something in the notion of the Body without Organs that can be understood from the perspective of the Buddhist *mahajana* better than from the perspective of ontology or psychoanalysis.

In *Méditation et Action*, Chogyam Trungpa says that there is nothing that corresponds to the word 'I' or to the sentence 'I am'. There are sensations, perceptions, impressions, imaginations and hypotheses that, in their permanence and in their becoming, we label with the word 'I'.

In their works, Deleuze and Guattari do not, to my knowledge, explicitly open the door toward Buddhist thought, but something in their thinking allows me to hazard the hypothesis of contiguity between rhizomatic thought and the thought of the void.

11
Chaosmosis

Between the end of the 1970s and the start of the 1980s, I happened several times to accompany Félix to the La Borde Clinic where he worked as a psychiatrist. La Borde is a short distance from Dhuizon where Félix lived. He had a house surrounded by trees, and it was there that he wrote, not in Paris, in the house on the rue de Condé crowded with friends, exiles passing through, and fascinating women. Sometimes Félix invited me to participate in the activities organized at La Borde by the guests at the clinic. I remember him asking me to hold a conference on free radios. There was a small crowd there to hear me, all seated on chairs arranged pell-mell, and I told Italian stories. At the end, a woman of a certain age took the floor to express her sympathy for Stalin and for comrade Togliatti, both of whom by then had been dead for quite a few years. At the same time, a Japanese dance troupe with painted faces, I forget the name, came to La Borde, and in the clinic's big courtyard they moved among the dismayed patients making improvised movements that cut through the air followed by long immobile suspensions.

*****

In an apparently simple and spontaneous way, Félix made connections between the traditionally separate territories of politics and psychoanalysis, of militancy and the cure. The connection point is the concept of the singularity of chaosmosis. And this connection point is singular, unrepeatable and different each time the connection occurs.

Guattari's contribution to the problematics of analysis and of the cure is called schizoanalysis. Schizoanalysis is not only a psychotherapeutic method, but also a way of conceiving of the relationship between singularity and its chaosmosis, that is, its conceptual, artistic, psychic, existential, political, relational and linguistic creations – the creations that constitute the world as a particular plane on which it is possible to start walking toward the encounter with other singularities.

The schizoanalytic conception is based on the idea that psychogenesis is singular, and thus therapy must be singular as well. This conception does not originate in the isolated field of psychotherapy as a result of a reflection on psychoanalysis, but belongs to a process of philosophical mutation that disperses models of solid thinking and introduces fluid thinking as a prevalent modality of understanding. Liquid, gaseous, psychochemical – and no longer mechanical – thinking.

By reviewing the formative lines of Guattari's thought, we can see through which concepts, thematic displacements and manipulations of word-tools he managed to put in movement weightless architectures of thought and therapeutic games of language.

'Happiness' (*felicità*) can seem a corny and banal word lacking philosophical content, an embarrassing word naming something impossible, an objective that we are all destined to miss. With this word, however, I intend to refer to the relationship between the singular drift and the cosmic game, because happiness and unhappiness are possible modes of this relationship. The harmony between singular drift and cosmic game is situated in the utopic point where identitarian obsessions and rigid refrains are dissolved. When this condition occurs, it is then possible to move with the lightness of someone who doesn't defend anything, someone who's not shackled by any identity or possession. This lightness means being the flow, not following it.

The weightless singularity does not presuppose any sense, nor does it discover any sense through its experience in the world. Sense unfolds as creation, as a connecting desire, as the delirium that illuminates the event. Schizoanalysis wants to make lightness possible, dissolving obsessions and rigid refrains through techniques aimed at displacing the focus of attention, through the proliferation of points from which semiotic flows, flows of worlds, can emanate.

If chaosmosis is the process through which singular creativity gives its own order to the chaotic constellations that circumscribe the conscious organism, then schizoanalysis proposes to act as an ecology of the mind and a therapy of becoming world.

From psychoanalysis to schizoanalysis

Psychoanalysis seems to be going through a profound crisis. In the issue of *Newsweek* published for New Year 2000, there was a list of things destined to disappear in the new century. The first would be psychoanalysis, while, according to the news weekly, psychic or esoteric magic would survive, made popular by entire armies of charlatans. For the better part of the twentieth century, psychoanalysis exercised an intellectual and therapeutic predominance as a form of interpretation and cure of neuroses. In the last few decades, we have witnessed a flourishing of the relational type of psychotherapy (like the one inspired by Gregory Bateson's thought) and of body-related psychotherapies such as bioenergetics. Finally, we have now seen the return of a refined and developed organicist therapy of psychopharmacology while on the horizon we can already discern the possibility of the decisive entry of genetics into the psychotherapeutic field.

The crisis of psychoanalysis does not correspond, in fact, to a quantitative diminution of psychopathologies, neuroses or mental suffering, quite to the contrary. To the extent that statistics and quantitative hypotheses can be taken seriously on issues concerning people's psychic well-being, common experience seems to suggest that mental illness and psychic suffering are penetrating into the folds of city life more pervasively or at least more evidently than ever.

Phenomena like depression and panic seem to have reached epidemic dimensions in the Western world. According to some observers, the crisis of psychoanalysis can be linked to a certain kind of hasty consumerism. The demand for a cure is ever more related to a demand for rapid consultations, and the spiritual consumer is ever less inclined to commit to the very lengthy timeframe required by the psychoanalytic cure. Prozac is better than anamnesis in the era of fast food and the acceleration of productivity.

*****

But this is not enough. Perhaps it is necessary to recognize that psychoanalysis is tied to an elitist status, not only because of its high costs and the lengthy times of treatment, but most of all because it requires an increasingly rare intellectual availability, today when mental time has become the principal source of economic value. For the members of the virtual class, who produce value by investing their cognitive labour, mental suffering is a sort of professional illness that is being cured with psychopharmacological means, tools of quick intervention aimed at restoring productivity to the mind. For the residual population excluded from the virtual circuit, psychopathology then becomes the new condition of normality: an aggressive normality, of which integralism, nationalism and conformism are common manifestations.

*****

Perhaps the crisis of psychoanalysis is truly derived from the normalization of psychopathology, from the fact that this is becoming a shared condition. How is it possible to maintain a specific area of analysis and psychotherapy when the processes of politics, production and communication seem to coincide with a progressive pathologization of social existence?

Naturally I realize how dangerous this position is.

In this way of thinking there is, first, a risk of flattening out individual histories, of reducing them to manifestations of a generalized social pathology and, secondly, we also run the risk of losing sight of the specificity of social analysis, definitively diluting it within a psychopathological frame. Finally, and more radically, one must ask: who gives us the right to judge as pathological the form of life that is emerging?

Here we have a problem that we must pose whenever we exercise judgment on the political, psychoanalytical or even aesthetic plane. What right do we have to consider as barbarous that which simply escapes our criterion for judgment? Is it perhaps not a question of an intrinsic limit of the human character in critical thinking? Is it perhaps not a question of a limit from which we cannot in fact be freed but that we should admit, simply recognizing our inability to understand what has already surpassed the limits of humanistic culture?

*****

When we are horrified when faced with the expanse of little houses decorated with lawn gnomes that are flourishing, for example, along the Sicilian coast, what right do we have to pretend that our taste is objectively superior to the taste of ninety per cent of Sicilians who like to construct Mexican-style houses with Nordic gnomes between the valley of temples and the Ionian Sea?

Is there not perhaps an irremediable contradiction between aesthetics and democracy?

And isn't it perhaps obvious that democracy is fated to win, screwing all ethics as well as humanistic and romantic aesthetics, finally imposing a new aesthetics made of Texas-style houses, lawn gnomes and fake-marble fences?[1]

In the behaviour of the vast majority of contemporary humanity, we must not see a superficial effect, as political ideologies or movements of opinion used to be. We must see the behaviour as the sign of an irreversible mutation that invests the social psyche and the very forms of cognition. We must see it as the effect of psychic, linguistic and cognitive automatisms that in no way can be contrasted or corrected by political action, civic edification or humanistic-humanitarian good sentiments. Human mutation simply makes humanistic universalism obsolete.

*****

Today, the ethnic conflicts generated by the displacements of mass populations, the political crises that result from them, the emergence of forms of racial, religious and national aggression seem to be retracing the path of events that in the first part of the twentieth century brought about the emergence of Nazi totalitarianism, the extermination of entire populations, and war. But what is happening today is a much larger extension of what happened in the 1920s and 1930s because the phenomenon concerns the vast majority of the world population. And in contrast with that era, it is not a question of a phenomenon provoked by the political will of fanatical groups.

Fanaticism has become a psychic automatism for the great majority of people. Psychopathology has become common sense.

*****

This is why the crisis of psychoanalysis may be understood as a decisive moment of passage. To get beyond this moment, I see the necessity of rethinking psychoanalysis as schizoanalysis, that is, an analysis that takes the singular and proliferating viewpoint of innumerable linguistic agents. In a certain sense, the cure for mental suffering and the interpretation of psycho-semiosis are becoming the new form of politics, the ultimate adequate politics in the productive post-industrial universe in which the mind is directly put to work and thus submitted to pathogenic regimes and rhythms.

From the crisis of psychoanalysis what emerges is schizopolitics, a politics understood as therapy of the collective functioning of language and as the dissemination of therapeutic flows in the circuit of social communication.

In the modern era, the political was the order imposed by rational will on the ensemble of imaginary, projective and communicative phenomena. Rationality – that is, the capacity to choose between alternative decisions according to a universal criterion – guided the will – that is, the capacity to pursue a plan and to impose it on the course of mental, communicative and material events. None of this exists any more. The signs that constitute the shared world of the Infosphere are too numerous and too quick to be critically analysed and understood. Consequently, the conditions no longer exist for a rational decision about their complex order, and they exist even less for a voluntary orientation of the course of global events.

*****

The course of global events thus presents itself as a psychopathological chaos in which individual and even collective minds lose any ability to guide their own destiny. Psychic suffering is derived from the feeling of rampant inadequacy between the pretext of a voluntary and rational finality of action and the reality of random and inconsistent movement, the reality of an imaginary deprived of order and hierarchies. It is the legacy of modern rationalism that produces a feeling of inadequacy in the collective mind.

*****

Schizoanalysis suggests another way of relating to politics. No possibility for rational selection exists, nor for voluntary finality. Schizoanalysis proposes to abolish any normative reference to rational judgment.

Schizoanalysis institutes the criterion for a singularity of judgment that is parallel to the singularity of communicative flows and existential aggregates.

No social coherence exists any more; rather, singular processes of socialization exist, and they can assume their own self-governance only if we are able to free ourselves completely from the normative and organizational reference to the social aggregate.

We must declare human society dissolved, because humanity no longer has any reason to remain together. The disappearance of universal judgment, of politics and of normativity does not indicate the provisional crisis of a rationality that a new universal rationality ought to succeed (as both dialectical historicism and progressivist political thought pretend). No human universality will ever exist anymore. Schizoanalysis starts from here.

Schizoanalysis is the ability to constitute sociality starting from the singular recombination of imaginative, mnestic and desiring elements.

The self-constitution of post-human singularities.

The reactive dynamics of territoriality

The pathology with which schizoanalysis concerns itself always arises at the meeting point between the ecology of the mind and the becoming of the world.

> Contemporary history is increasingly dominated by rising demands for subjective singularity – quarrels over language, autonomist demands, issues of nationalism and of the nation, which, in total ambiguity, express on the one hand an aspiration for national liberation, but also manifest themselves in what I would call conservative reterritorializations of subjectivity. A certain universal representation of subjectivity, incarnated by capitalist colonization in both East and West, has gone bankrupt – although it's not yet possible to fully measure the scale of such a failure. (Guattari, *Chaosmosis*, 3)

In the 1990s, the psychochemical constitution of the planet entered into a dark phase of turbulence. The molecular substances that were formed in the course of the rapid maceration of modernity unleashed by '68 kept compounding, composing and interacting, finally resulting in a phase that we could consider as a de-volution. The crisis of industrial labour, the collapse of the authoritarian Socialist bloc, the digital revolution, the disaggregation of the patriarchal family, the diffusion of a movement of feminist autonomy that upsets masculine power, the mediatization of planetary communication, and all the innumerable local processes of mutation are provoking a reactionary recouping. Guattari described the aggregate of these processes by talking about the contradictory and psychotic dynamic of the movement of deterritorialization and the counter-movement of reterritorialization.

Let us pause here with the synthesis of these concepts offered by Vincent Descombes:

> According to the *Anti-Oedipus*, universal history is a process of 'deterritorialization'. 'Deterritorialization' defines the essential movement of capitalism. Capitalism emerges at the end of history, and is the 'universal truth' of history. But what exactly is 'deterritorialization'? It is the transition from a *coding* to a *decoding*. Here the term 'coding' does not refer to a linguistic operation ... but to the way that society regulates production (which, as we will recall, intends both the Marxist 'social production' and the 'desiring-production') ... Capitalism originates therefore in a generalized decoding ... Ancient ritual, ceremony, every form which was once respected or held *sacred* must disappear. Capitalism is defined as a 'cynical' system which, in order to function, needs no appeal to the sacred, to belief. We are confronted here with the same aberration as before. The product of cultural training was to have been the sovereign individual, but is in fact the man of negativity. In the same way, capitalism, as defined by the cynicism of decoding, was to have brought liberation, since it destroyed all the beliefs and prohibitions which had enthralled humanity; but the reality of capitalism ... is the greatest repression of desiring-production ever witnessed in history. By destroying all *ties*, capitalism should have created the conditions for the blissful nomadism of a detached and *absolute* individual, as a consequence of 'deterritorialization'. However, in this liberation of all flux, it has produced a world of nightmare and anxiety. Why should history have failed? The reason is that 'deterritorialization' is accompanied by a perpetual 'reterritorialization'. Capitalism postpones the limit towards which it tends (nomadism) by restoring artificial 'territorialities' (beliefs, forms). 'Everything returns or reappears – states, fatherlands, families.' (*Modern French Philosophy*, 176–7, emphasis in the original; citation from *Anti-Oedipus*, 34)

On the one hand, a general deterritorialization carrying enormous potentialities, conditions of enrichment of experience, of collective and individual bliss but also a frightful imbalance, a feeling of loss and dispossession, a desperate need for identity. On the other hand, a movement of aggressive reterritorialization, the obsessive affirmation of identity, the blinding prevalence of the feeling of belonging, the conformist cancellation of singularity.

On the social plane: the workers' struggles of the 1960s and 1970s have led capital toward a general technological restructuring that constitutes an enormous displacement of productivity from the human workforce toward electronic automation. This provokes the liberation of increasing amounts of labour time. This is an effect of deterritorialization to which the 1990s culture (both capitalist and labour-centred) is unable to react except in the form of panic and anguish. We have not succeeded in outlining the politics of redistributing the work schedule, nor in undertaking processes for the liberation of social time. Consequently society has not succeeded in semiotizing the reduction of labour time except in terms of unemployment, impoverishment and marginalization. This has unleashed a powerful movement of reterritorialization, reactionary and impotent, but nonetheless enraged. The workers' struggles of the 1980s (the occupation of Fiat in autumn 1980, the Union Miners' struggle in 1983, the agitation of French and Italian steelworkers) were all losing battles because they barricaded themselves in the position of defending forms of labour which are obsolete, useless, difficult and depressing all at once.

On the ethnic and cultural plane: the great global migration from the South to the North that began in the 1980s and 1990s produced an effect of cultural, aesthetic and sexual contamination that enormously enriched the possibilities of experience, production and exchange. But this movement of deterritorialization provoked reactions of dismay and disorientation in the majority of the population. This changed the urban landscape, behaviours, semantic relations, the rituals of daily life, and this gave rise to desperate reactions: the reaffirmation of an identity that is just as imaginary and fictitious as any other originary claim. But it is precisely the fragility and the illusory artificiality of this claim that mobilized aggressive nationalist and racist energies.

> Sociology, economic science, political science and legal studies appear poorly equipped to account for the mixture of archaic attachments to cultural traditions that nonetheless aspire to the technological and scientific modernity characterizing the contemporary subjective cocktail. Traditional psychoanalysis, for its part, is hardly better placed to confront these problems, due to its habit of reducing social facts to psychological mechanisms. In such conditions it appears opportune to forge a more transversalist conception of subjectivity, one which would permit us to understand both its idiosyncratic territorialized couplings (Existential Territories) and its opening onto value systems (Incorporeal Universes) with their social and cultural implications. (*Chaosmosis*, 3–4)

On the techno-communicational plane, the diffusion of digital technologies opens unthinkable possibilities for connections, the expansion of the sphere of relations, and the quantity of available information. At the same time, social relations are becoming virtual; the relation between bodies in space is partially being substituted by relations between disembodied speakers, semiotic actors without immediate social contact. Virtual deterritorialization, the transfer of the locus of political decisions in a sphere that has no contacts with social daily life, produces violent reactions from those members of residual (but not minoritarian) sociality who feel themselves disenfranchised and marginalized. In response, effects of massive reterritorialization are unleashed and become manifested with the reemergence of strongly territorial identities.

There is a profound relationship, but one difficult to decipher, between communicational globalization, cultural homogenization, the generalized transparency of forms of life, and the aggressive reaffirmation of tribal, ethnic and religious forms of identity that seemed buried in archaism. On the psychic plane, phenomena of informational overload act as a cause of anxiety, panic, and as a disturbance of the concrete emotional sphere. The background noise produced by the proliferation of informational emissions of every kind overpowers the voice of the father as head of the family, who feels dispossessed of his sole, miserable power, as well as the voice of the lover, who feels overwhelmed by other fantasmatic, distant lovers.

*****

Guattari was not able to see the unfolding of the telematic web as a planetary anthropological phenomenon, since the year of his death was also the year in which the World Wide Web began its diffusion. But he had foreseen all its possibilities and potential consequences, and had conceptually anticipated its dynamics.

> Should we keep the semiotic productions of the mass media, informatics, telematics and robotics separate from psychological subjectivity? I don't think so. Just as social machines can be grouped under the general title of Collective Equipment, technological machines of information and communication operate at the heart of human subjectivity, not only within its memory and intelligence, but within its sensibility, affects and unconscious fantasms. Recognition of these machinic dimensions of subjectivation leads us to insist, in our attempt at redefinition, on the hetero-

> geneity of the components leading to the production of subjectivity. (*Chaosmosis*, 4)

The process of subjectivation passes through all the effects of deterritorialization and reterritorialization that are born from communication and depend on technologies that enlarge, complicate and channel communication. Here we are at the heart of a central question in Guattari's thought, the process of subjectivation. This concept has a precise sense: a universal form of the Subject does not exist, neither in the historicist sense that Marxism inherited from Hegel, nor in the Freudian sense of a universal topology of the Unconscious.

The unconscious is a factory, not a theatre. It is a site of singular production, of machining processes that install singular elements following a rule of singular functioning. Thus one must not speak of the subject, but of subjectivation. And subjectivation depends on the webs that weave the singularity into its relations with singular flows.

*****

The notion of subjectivity is ambiguous because it is located at the crossing point from which a perspective of historicist finality, inherited from the heavy thinking anchored to the dominant categories of the Hegelian matrix, and a proliferating, light and creative one start to diverge.

One must be clear about this word. Guattari did not talk about subjectivity, but about the production of subjectivity: this subjectivity does not pre-exist the process of production or the psychic, aesthetic and ideological flows that traverse it, seep through it, colour and displace it. It does not pre-exist the anticipations, the desires and the fears that shape its form.

The context within and against which Guattari elaborated his conception of the production of subjectivity is that of Freudian and Lacanian structuralism. This was a model of thought that pretended to locate in the phenomenology of existence and desire the manifestations of a structuring psyche, of Oedipus, the unfolding of guilt and remorse, the sudden emergence of a repressed past which remains inscribed in the flow of language.

Guattari began his philosophical and psychoanalytical trajectory precisely by distancing himself from this structuralism and Oedipalism.

> The same movement towards a polyphonic and heterogeneous comprehension of subjectivity leads us to consider certain aspects of

> contemporary research into ethology and ecology. Daniel Stern, in *The Interpersonal World of the Infant*, has notably explored the preverbal subjective formations of infants. He shows that these are not at all a matter of 'stages' in the Freudian sense, but of levels of subjectivation which maintain themselves in parallel throughout life. He thus rejects the overrated psychogenesis of Freudian complexes, which have been presented as the structural 'Universals' of subjectivity. Furthermore, he emphasizes the inherently trans-subjective character of an infant's early experiences, which do not dissociate the feeling of self from the feeling of the other. A dialectic between 'sharable affects' and 'non-sharable affects' thus structures the emergent phases of subjectivity. A nascent subjectivity, which we will continually find in dreams, *délire*, creative exaltation, or the feeling of love ... (*Chaosmosis*, 6)

Guattari here traces a conception of psychogenesis which is not structured, like the Freudian one, but singularizing.

In Freud, psychogenesis is represented as the manifestations of psychic complexes preconstituted by the play of impulses and forms of repression (and Lacan formalized this play finally by representing it as a *mathème*, as a linguistic game, rather than as the magma of psychic organicism).

Guattari eventualized psychogenesis: psychogenesis is this becoming singular, and the word 'become' must then be used in the plural ('becomings') because becomings are projected universes emanating from singular psychogeneses. This does not, in fact, mean that one has to deny any cognitive and practical value to Freudian discourse. Instead, rather than consider it as a description of the universal human psyche, Guattari proposes to see it as a creation, as a modality of semiotization of the unconscious world, as a mythological tale that shapes and illuminates at the same time.

> In the same way that Christians invented a new form of subjectivation (courtly chivalry and romanticism, a new love, a new nature) and Bolshevism a new sense of class, the various Freudian sects have secreted new ways of experiencing – or even of producing – hysteria, infantile neurosis, psychosis, family conflict, the reading of myths, etc. (*Chaosmosis*, 10)

The Freudian model is not understood here as an imperatively universal model, but as a model of semiotization of intra-individual life, a paradigm for the production of subjectivity that must be anthropologically situated. It is one of the fundamental passages for understanding Guattari's thought as a critique of structuralism, in psychoanalysis as well as in linguistics. Guattari did not, in fact, discard the Freudian notion of the Unconscious, but he situated it both in the context that produces it and in the one that it itself produces. The Oedipal complex, says Guattari, is not a universal interpretive category of human psychogenesis, but a model for the semiotization of familial, affective and erotic relations which is precisely identifiable in the context of the modern bourgeois family and of the monotheistic – Judaic, Christian, Protestant – sentiment of guilt.

> When a Bahun child, reared on the purity of his caste, discovers a father or close relative privately violating such behaviour, then he comes to share a deep and wicked secret about high caste life, that it masks a certain degree of fraudulence, that the high ideals of purity are often something of a public façade; and the sense of superiority is largely undeserved. In sharing this secret pollution of the Bahuns as a caste, the individual then develops a secondary concern, that it will be found out. Unlike the non-Bahun child, then, the Bahun does develop a certain inner psychological dynamic that involves guilt. Rather than motivating responsible behaviour, however, one of the consequences of this guilt is a deep and widespread paranoia, that has adverse effects in high caste social organization. (Dor Bahadur Bista, *Fatalism and Development*, 73)

The obsession with purity is the origin of the sense of guilt, and the sense of guilt puts in motion the classical mechanisms of psychoanalysis, the chain of the double bind implicit in prohibition and transgression. But when we speak about psychopathology, we must carefully avoid defining the description of universal categories. In a culture based on the respect of caste purity, the originary scene will not, in fact, be Oedipus, but the socially impure touch. It is from this that a tear is made in the apparently tightly woven fabric of Being. This feeling of inadequacy and the contrast between the interdiction cast by purity (the tightly woven fabric of Being) and its inadmissible transgression or an impure desire (the existential leak) does not function according to universal modalities and categories, but according to anthropologically determined modalities.

In Dor Bahadur Bista's book, one can locate other useful elements for a similar comparative analysis of the anthropological contexts of psychogenetic processes: mixed marriage, for example (one between a man of the Brahmin caste and a woman of the Kastri caste) can put in motion chains of interdictions and impurities with the highest pathogenic potential. But pathogenic processes must then be studied in a specific situation, through a method of singularization, and not at all through a method of psychogenetic universalization.

Guattari synthesized the schizoanalytical method in this way:

> My perspective involved shifting the human and social sciences from scientific paradigms toward ethico-aesthetic paradigms. It's no longer a question of determining whether the Freudian Unconscious or the Lacanian Unconscious provides scientific answers to the problems of the psyche. From now on these models, along with the others, will only be considered in terms of the production of subjectivity – inseparable as much from the technical and institutional apparatuses which promote it as from their impact on psychiatry, university teaching or the mass media ... In a more general way, one has to admit that every individual and social group conveys its own system of modelising subjectivity; that is, a certain cartography – composed of cognitive references as well as mythical, ritual and symptomatological references – with which it positions itself in relation to its affects and anguishes, and attempts to manage its inhibitions and drives. Psychoanalytic treatment confronts us with a multiplicity of cartographies. (*Chaosmosis,* 10–11)

The process of the cure cannot be understood (as familializing psychoanalysis or normalizing psychiatry does) as a reduction of the deviant psyche to behavioural, linguistic and psychic norms recognized by society. It must be understood, on the contrary, as the creation of psychic nuclei able to make habitable a certain psychic cartography, as a singularization that can be happy (*felice*) in itself. This is the task of schizoanalysis: to follow delirium in order to render it coherent and sharable, to open it to friendship in oneself and to friendship with the other. To dissolve the identitarian nuclei that petrify the refrain, to assemble the refrain with other refrains. To reopen the channel of communication between the individual drift and the cosmic game.

Linguistic nomadism

Guattari's thought can be described as a strategy of singularization of the psychogenetic process, but also in the same way, as a strategy of singularization of the language act. In *The Machinic Unconscious,* Guattari squares accounts with linguistic structuralism and begins to displace the problematic field toward the domain of pragmatics.

> There is nothing less logical, less mathematical, than a language. Its 'structure' results from the petrifaction of a kind of catch-all from which its elements emerge by borrowings, amalgams, agglutinations, and misunderstandings ... It is the same with linguistic laws as with anthropological laws, for example those that concern the incest taboo: seen from the distance of a grammarian or an ethnologist, they seem to have a certain coherence, but when one gets closer to them, everything gets confused, and one notices that it's a question of systems of arrangements able to be drawn in all directions and of rules able to be used in all sorts of ways. (*L'Inconscient machinique,* 25, our translation)

The Chomskyan definition of rules of the functioning of language, of universal structures of linguistic competence appears to be generalizations unable to explain the real becoming of linguistic communication, because linguistic communication functions in relation to non-linguistic contextualized levels of communication that continually redefine language. Any individual, Guattari observed, continually passes from one language to another. The ideal speaker that Chomsky proposes is an abstraction deprived of analytic value when one tries to examine its real language production.

> There is no universality of language, nor is there any universality in language acts. Each sequence of linguistic expression is associated with a network of semiotic links of all kinds (perceptive, mimetic, gestural, image-thoughts, etc. ...). Any signifying statement crystallizes a silent dance of intensities played out both on the social body and on the individuated body. From language to speaking in tongues [*glossolalie*], all transitions are possible. (*L'Inconscient machinique,* 31, our translation)

Claude Hagège usefully wrote on the notion of linguistic universals in order to set up a methodological question, the question of the relationship between linguistic science and language as a real process.

> Formal universals are not, in fact, language universals, but rather, general conditions of coherence for linguistics. They are epistemological

> requirements which can teach us something about logical systems, about methods to be used in the social sciences and about the ingenuity of the people who develop them; they tell us little about actual languages ... or about the species characterized, at least in part, by those languages. We must not, simply because a linguistic theory requires certain methodological procedures, confound these procedures with the object to which they are applied. (Hagège, *The Dialogic Species*, 35)

And Hagège also observed that:

> Linguistics, born of the recognition of constants, is in large part becoming a study of variation against a background of invariants; a science that no longer contemplates constants in and of themselves, but subsumes them under the thousand faces of their variants. (67)

This consideration coincides with what Guattari wrote about language in *L'Inconscient machinique*. Rather than question language from the point of view of structure, Guattari questions it from the point of view of eventuality, of the speech act: removed from the generative conception of structuralism that makes of language the execution (that is, repetition) of a preconstituted and innate programme, language is thus rethought according to a nomadic vision.

Language is essentially deterritorialization, the invention of a context. We generally say that the meaning of a statement depends on the context, but we must add that the meaning of the context in turn depends on the statements that intersect it. As the invention of contexts, the constant mutation of laws within which the game is played, language is stateless and nomadic. Which does not exclude, obviously, identifying nuclei – truly anchored, territorialized – of linguistic flows from being determined. The stabilization of national languages, the formation of codes, subcultures and so forth constitute regimes of linguistic territorialization. In this area, communication does not function as redefinition and invention of contexts, but as its confirmation, as identification and self-identification.

And here we turn to the theme of the refrain.

The refrain is above all a musical phrase that returns in the course of a song: a phrase that, in returning, constructs and brings forth the complex rhythm. It is a factor of assemblage: by constructing rhythm and bringing forth the song's complex rhythm, the refrain functions as the structuring element in language, in existential behaviour, and in history.

> A child that sings in the night because of his fear of the dark tries to reestablish control of events that are too quickly deterritorializing for his liking and that begin to proliferate in the cosmos and in the imaginary. Each individual, each group, each nation thus equips themselves with a range of basic refrains for conjuring. (*L'Inconscient machinique*, 109, our translation)[2]

The refrain is an obsessive ritual that is initiated in linguistic, sexual, social, productive, existential behaviour to allow the individual – the conscious organism in continuous variation – to find identification points, that is, to territorialize oneself and to represent oneself in relation to the world that surrounds it. The refrain is the modality of semiotization that allows an individual (a group, a people, a nation, a subculture) to receive and project the world according to reproducible and communicable formats.

In order for the cosmic, social and molecular universe to be filtered through an individual perception, in order for it, we may thus say, to enter the mind, filters or models of semiotization must act, and these are models that Guattari called refrains. The perception of time by a society, a culture or a person is also the model of a truly temporal refrain, that is, of particular rhythmic modulations that function as modules for accessing, awaiting and participating in cosmic temporal becoming.

> From this perspective, universal time appears to be no more than a hypothetical projection, a time of generalized equivalence, a 'flattened' capitalistic time; what is important are these partial modules of temporalization, operating in diverse domains (biological, ethological, socio-cultural, machinic, cosmic ...) , and out of which complex refrains constitute highly relative existential synchronies. (*Chaosmosis,* 16)

What is the fundamental passage through which the anthropological transformation of modern capitalism is determined? This passage consists in the creation of refrains of temporal perception that invade and discipline all society: the refrain of factory work, the refrain of working hours, the refrain of the salary, the refrain of the production line. The postindustrial transition brings along with it the formation and imposition of new refrains: the refrain of electronic speed, the refrain of information overload, the refrain of digitalization.

> My feeling of personal identity is thus pulled in different directions. How can I maintain a relative sense of unicity, despite the diversity of components of subjectivation that pass through me? It's a question of the refrain that fixes me in front of the screen, henceforth constituted as a projective existential node. My identity has become that of the speaker, the person who speaks from the television. (*Chaosmosis*, 16–17)

In communication, obsessive and fixated types of nuclei are determined; certain refrains thicken and solidify, entering into resonance and producing effects of double bind. When the existential flow gets rigidly brought back to logical, mythological, ideological or psychic refrains, behaviour tends to become paranoid. For example, when the money refrain becomes the structuring element of all social and communicative life, this engenders behavioural paradoxes, paranoid anticipations, social double binds, and depression.

In the relation between the fluidity of social and communicational existence and the fixation on obsessive refrains, the psychopathologies of collective life are created.

> In the case of neurotic identity, sometimes the refrain develops into a 'hardened' representation, for example an obsessive ritual. (*Chaosmosis*, 17)

In the book's subsequent pages, Guattari reverses the analyses of the hardening of refrains from a therapeutic perspective. What does the task of the therapist or the analyst consist of?

> Analysis is no longer the transferential interpretation of symptoms as a function of preexisting, latent content, but the invention of new analytic nuclei capable of bifurcating existence. A singularity, a rupture of sense, a cut, a fragmentation, the detachment of a semiotic content – in a Dadaist or surrealist manner – can originate mutant nuclei of subjectivation. (*Chaosmosis*, 18)

The therapeutic problematic is therefore placed at the junction between the refrain's signifying hardening, that is, the succession of particle-signs that proliferate in the psychochemistry of the Infosphere, and the desiring linguistic drift, delirium as an a-signifying trajectory of words, sounds, writing, images, the body – in short, the sign. And therapy consists in liberating a-signifying flows because they create new assemblages and eventually put in motion unforeseen signifying chains that did not succeed in manifesting themselves before because they were blocked by dominant assemblages of meaning, by obsessions, by social and linguistic laws, and by communicative double binds that generated neuroses and the pressure to repeat.

> Schizoanalysis obviously does not consist in miming schizophrenia, but in crossing, like it, the barriers of non-sense which prohibit access to a-signifying nuclei of subjectivation, the only way to shift petrified systems of modelization. (*Chaosmosis*, 68)

Bridges over the abyss

Guattari's thought provoked a change in the image of thought and in the function that we attribute to philosophy. On the one hand, the planes intersect in such a way as not to allow any separation between schizoanalytic, ecological, political and philosophical planes. On the other hand (and here the discussion leads us inevitably to Gilles Deleuze and their work in common), we can no longer speak of philosophy if we do not speak about the eventual abyss over which the philosopher, this happy sage, walks.

What is the eventual abyss? It is the full awareness of the non-being of sense, of the ontological non-foundation of sense. And at the same time, it is the consciousness of sense's purely eventual, conjectural and circumstantial character as the projection of a world that is alive insofar as it is a world of experience, of irradiation, of relations, of friendships, love, projects, promises and challenges. Over the eventual abyss stretch weak bridges: the bridges of philosophy, that is, of friendship. This is the discourse that Deleuze and Guattari deploy in their final book, *What Is Philosophy?*

In the introduction, 'The Question Then ...', Gilles and Félix question the sense of the anti-universalist, anti-structuralist, and anti-historicist philosophy that was outlined through their work.

> The question of *what is philosophy?* can perhaps be posed only late in life, with the arrival of old age and the time for speaking concretely ... [Those who asked the question before] were not sober enough. There was too much desire to *do* philosophy to wonder

> what it was ... That point of non-style where one can finally say, 'What is it I have been doing all my life?' had not been reached. There are times when old age produces not eternal youth but a sovereign freedom, a pure necessity in which one enjoys a moment of grace between life and death, and in which all the parts of the machine come together to send into the future a feature that cuts across all ages. (*What Is Philosophy?*, 1–2, emphasis in the original)

In this point (of non-style) it becomes possible to hang over the abyss and define philosophy in relation to the abyss. 'Philosophy is the art of forming, inventing, and fabricating concepts' (*What Is Philosophy?*, 2). The essential centre of philosophical activity is inventing concepts; but then what are concepts?

> The concept is an incorporeal, even though it is incarnated or effectuated in bodies. But, in fact, it is not mixed up with the state of affairs in which it is effectuated. It does not have spatiotemporal coordinates, only intensive ordinates. It has no energy, only intensities; it is anenergetic (energy is not intensity but rather the way in which the latter is deployed and nullified in an extensive state of affairs). The concept speaks the event, not the essence or the thing – pure Event, a hecceity, an entity: the event of the Other or of the face ... The concept is defined by *the inseparability of a finite number of heterogeneous components traversed by a point of absolute survey at infinite speed.* (*What Is Philosophy?*, 21, emphasis in the original)

This is why the philosopher has no real desire to argue or debate. Philosophers despise the confrontation of ideas, thoroughly avoiding it (and in this, they differ from the swarm of pastry vendors crowding journalistic, televisual and cultural debates). Because there is nothing to debate or argue about when one discusses philosophy. It is not a question of comparing concept to concept, of seeing what is true, how adequate this or that concept might be, and so forth. It is a question of initiating a plane of immanence, of constructing a new perspective, of opening a gap in the blindness of common vision, precisely of constructing bridges over the abyss. And it is a question of walking across these bridges accompanied by whoever shares the intensity that allowed it to be constructed.

> The philosopher is the concept's friend; he is potentiality of the concept. That is, philosophy is not a simple art of forming, inventing, or

> fabricating concepts ... More rigorously, philosophy is the discipline that involves *creating* concepts. Does this mean that the friend is friend of his own concepts? Or is the actuality of the concept due to the potential of the friend, in the unity of creator and his double? The object of philosophy is to create concepts that are always new. (*What Is Philosophy?*, 5, emphasis in the original)

But what does the friend mean, or rather the lover? We can say that love and friendship have the same power of the concept, the capacity of stretching bridges over the abyss, of allowing the wanderer to cross the abyss. There is nothing in love, nothing in friendship; love is not the truth of anything, it does not prove anything, does not demonstrate anything, does not 'mean', does not signify.

Thus it is not debated, but flies over the abyss of being.

Is this not perhaps the same task of the analyst, of therapy, if we cease considering it as the regaining of a lost normality? Reality addicts pretend that a normal behaviour exists, a normal language, a normal world. This is not true. Normality has no basis on which to affirm its right to judgment.

There is no fullness to which one must conform, but rather an abyss over which we must soar. Culture is the invention of techniques to soar over the abyss. Friendship is their form.

> Isn't the task of analysis precisely to recharge Expression with semiotic heterogeneity and to run counter to the disenchantment, demystification and depoetisation of the contemporary world? (*Chaosmosis*, 76)

12
The Provisional Eternity of Friendship

What is philosophy?

Twenty years after *Anti-Oedipus* and ten years after *A Thousand Plateaus* there appeared *What Is Philosophy?*, a book that can be considered as the recognition of the conceptual backstreets, of the tool boxes that have been used to construct the internal fortress of language, as pulling back the curtain on the backstage scene.

Deleuze and Guattari's final book places all the cards on the table; it is their least mysterious book, going straight to the heart of the problem that is posed precisely in the book's title: what is philosophy?

Philosophy is an activity of creation; it creates concepts, and concepts are a kind of crossing bridge over the abyss of chaos. Only friends, only people who understand each other without needing to debate, refute, agree with each other, can walk together across this bridge.

There is no project of reaching the truth, of realization, behind the conceptual creation that Gilles and Félix are talking about.

Philosophy does not abolish chaos, nor does it subject it to the definitive order of reason. Chaos is always first and last; it is the sea in which conceptual creation swims.

> A concept is a set of inseparable variations that is produced or constructed on a plane of immanence insofar as the latter crosscuts the chaotic variability and gives it consistency (reality). A concept is therefore a chaoid state par excellence; it refers back to a chaos rendered consistent, become Thought, mental chaosmos. And what would *thinking* be if it did not constantly confront chaos? ... We call *Chaoids* the realities produced on the planes that cut through the

> chaos in different ways. (*What Is Philosophy?*, 208, emphasis in the original)

The activity of connecting – thought, creation, movement – should not be conceived as the instauration of an order. The activity of connecting does not presuppose any pre-existing connection, nor should it be proposed as the finality of an optimal, rational connection, nor as consensus.

In effect, the activity of connecting is the desire, the condition of an itinerary that is invested with sense solely for whoever undertakes the trip.

> What does *friend* mean when it becomes a conceptual persona, or a condition for the exercise of thought? Or rather, are we not talking of the lover? Does not the friend reintroduce into thought a vital relationship with the Other that was supposed to have been excluded from pure thought? (*What Is Philosophy?*, 3–4, emphasis in the original)

The pretext of constructing thought as a rigorous exercise of formal rules, the naïve illusion of logical reduction does not understand the essential aspect of philosophical thought, which consists in following the line of infinity, in translating into concepts the uninterrupted activity of conceptualizing the relationship with chaos. Pure thought, from Kant on to logical positivism, deluded itself in establishing a pure constitution of the *I* beyond the tumultuous, unstoppable, material, physical and neurological relations between the brain and the circumstantial world.

> Chaos makes chaotic and undoes every consistency in the infinite. The problem of philosophy is to acquire a consistency without losing the infinite into which thought plunges (in this respect chaos has as much a mental as a physical existence). To give consistency without losing anything of the infinite... (*What Is Philosophy?*, 42)

The philosopher's activity consists in tracing the signs that might give consistency to the surrounding chaotic constellations (physical, psychic, political, geographical, relational, linguistic, aesthetic, emotional). Concepts are these signs, and they allow movement within the chaos of worldly and infrapsychic objects as if these had an order. Order is a projection of chaosmosis that is realized through the chaoid activity *par excellence*, that of creation: creation of concepts, creation of artistic, narrative, poetic and musical assemblages.

These chaoids have the function of making of the world a circumstantial world (*Umwelt*), a populated and inhabited world. However, this order is absolutely provisional, just as friendship is provisional, allowing different actors to recognize each other in the same refrain, to follow the same rules, to breathe with the same rhythm.

> The first philosophers are those who institute a plane of immanence like a sieve stretched over the chaos. In this sense they contrast with sages, who are religious personae, priests, because they conceive of the institution of an always transcendent order imposed from outside by a great despot or by one god higher than the others ... Only friends can set out a plane of immanence as a ground from which idols have been cleared. (*What Is Philosophy?*, 43)

Infinity is the continuous unmaking of affective-chaosmotic orders and the continuous creation that constitutes them in new and always mutable planes of consistency. It's like a competition, a game of tag between mind and brain, between chaos and chaoids, between conceptual refrains and infinity. It's a problem of speed: the constellations of which the world is composed are transformed at mad speed, and the world proliferates in all directions along continually broken lines. Moreover, the brain, as neurophysical activity, as creative potential, moves at a crazy speed. Then there is the slowness of singularization, of assemblages between singular, conscious organisms and infinite proliferation. The chaoids function like reductions of complexity, like grids that allow the organism to recognize the world in which it moves, to interact with it with full awareness.

But we do not always succeed in maintaining this speed, and the game can generate a collapse, a loss of rhythm, a painful sensation of disorder, an anxiety and a feeling of panic and also depression, that is, a disconnection from the chaosmotic rhythm and the renunciation of desiring investments.

> Chaos is defined not so much by its disorder as by the infinite speed with which every form taking shape in it vanishes. It is a void that is not a nothingness but a *virtual*, containing all possible particles and drawing out all possible forms, which spring up only to disappear immediately without consistency or reference, without con-

> sequence. Chaos is an infinite speed of birth and disappearance. (*What Is Philosophy?*, 118, emphasis in the original)

The thicker the infosphere, and the more intense and rapid the proliferation, the more the mind risks drowning in the disorganized potentiality of the brain, and the more the danger grows that the rhythm will be broken.

> We require just a little order to protect us from chaos. Nothing is more distressing than a thought that escapes itself, than ideas that fly off, that disappear hardly formed, already eroded by forgetfulness or precipitated into others that we no longer master. (*What Is Philosophy?*, 201)

Philosophy, like poetry, like schizoanalysis, are creative activities, in the sense that they do not have to restore a pre-existing order (Platonic truth or normative psychic equilibrium or the originary order of things). They undertake a struggle against chaos that is infinite, just as chaos is infinite. 'The struggle against chaos does not take place without an affinity with its enemy,' Gilles and Félix say. (*What Is Philosophy?*, 191)

Irony, play, language, history, singularity

There also is a discourse here on politics. Politics aims at institutionalizing chaoids, that is, at stabilizing in time the forms that are born within social and communicative becoming.

The history of civilization is the history of this chaoid petrification within states, churches, political parties and institutions of all kinds.

During the modern era, this process of fixation of becomings, of making nomadism sedentary, has reached the fullest intensity and efficiency.

States are profoundly inserted into the flesh, blood and mind of populations. Massacres, genocides, deportations have been permanently institutionalized within the chaoid-State.

The capitalist economy has brought strength and equilibrium to politics, but at the same time, it has produced movements of deterritorialization that break the equilibrium and deterritorialize populations, forms of life, culture, and thus disturb institutions producing a continuous crisis of the State. Wars in the twentieth century and the planetary state of civil war of this desperate end of the century are the sign of an irreversible crisis in relations between capitalist economy and politics.

During the modern era, complexity has been reduced to the practical and epistemic categories of political governance. We could say that the speed of proliferation of socio-communicational chaos has been governed by the conscious elaboration of chaoid institutions on the part of organized human will.

This capacity to govern has exhausted itself over the course of our century.

Social complexity is no longer governable and not even comprehensible by politics.

But to this, political thought has reacted in the course of the century with aggressive movements.

The more fragile its practical hold on the world has become, the more politics has tried to maintain its unlimited domination.

1968 is the culminating moment of this delirium of omnipotence of the declining politician, but at the same time, it is the comprehension (torn, fragmented, confused) of another kind of being of thought: one founded on friendship, on nomadic and provisional community. 1968 inaugurated the coming to consciousness of the ironic character of thought, that is (since the two are the same thing) of chaosmotic becoming. When thought wants to impose a world form by creating ideology, it institutes a principle of seriousness that substitutes the integralist pretext of truth for the play of communication.

Irony is the awareness of the playfulness of language and of the singular (but not therefore weak) character of the ethical perspective instituted by the play of language.

1968 is a watershed in the history of thought because it is then that the dialectical illusion of modern thought comes to its end.

But at the same time, in its ideological variant, 1968 conceived of the movement as a moment of realization of thought. Thought became ideology insofar as it wanted to impose the universal form of the world, reducing difference to identity. At first, continuing the Enlightenment enterprise, the twentieth century tried to impose universal identity (socialism, democracy or the market). When this attempt to impose universal principles collapsed (and 1989 is the moment when this happened), it was then that the ideology closed itself into the compound of particular identities, of reemerging nationalisms, and widespread forms of integralism.

Particular identities have nothing to do with difference. Difference is becoming other, infinite play that unfolds along the edge between chaos and creation, chaosmosis.

Difference is the singularity of worlds that are derived from creative projection, and that are only consolidated in friendship, in sharing desire.

Only lovers know difference.

But from within this movement (which for the sake of simplicity we have defined as 1968) was also born the consciousness of an irreducibility of the processes to any historical totalization. *Anti-Oedipus* represents the passageway to this consciousness.

From the failure of the dialectical and totalizing aspirations of 1968 emerged a path of disillusionment that was manifested through Heideggerism.

These philosophies that in the 70s and 80s (in different ways, Derrida and Vattimo) reintroduced Nietzsche and Heidegger had understood thought as dialogical activity, as infinite retreat within the web of interpretations.

In this way, language institutes the sayable of the world but does not constitute any dimension able to sustain a path, neither of an historical nor an existential type.

'Weak thinking' [*il pensiero debole*] freed itself from the totalitarian pretext of founding the historicity of the world, but lacked the goal of projecting a plane of consistency. Weak thinking does not succeed in conceptualizing singularity.

But precisely because it is incapable of conceptualizing singularity, of assembling it in friendship, in desire, the world at the end of the century fell back into the barbarity of identity.

The need for belonging and for identitarian bases followed the panic unleashed by the postmodern condition. This was because from 1989 no transparency resulted, because the result was the reemergence and valorization of opacities, that is, of integralist particularisms.

Deleuze and Guattari's thought, on the contrary, proposes an ethical desiring singularity.

What founds communication is desire. Friendship is a desiring play that is able to be singular.

Historical-dialectical thought proposed an imperious seriousness of enunciation – language that pretends to act on the world in a totalizing way and institutes in effect a violent totality.

The desiring thought of Gilles and Félix conceives of the pathway as sharing in lightness. Enunciation is not guaranteed by any truth, nor can it be imposed, nor can it institute a world that is imposed by other enunciations.

The desiring assemblage always emits ironic enunciations, the bridge over which one can walk only if one shares an intensity.

It is said that in the period of the First World War, in a bar in Zurich, Vladimir Illich Lenin and Tristan Tzara met, without ever having associated before.

The language of Lenin tried to create the world with the strength of the will, of law, of power.

Tzara used language as irony, as the creation of worlds in which will, law and power were suspended.

If they had understood each other, the twentieth century would have been lighter. If they had been friends, they would have undertaken the construction of small crafts able to navigate on the ocean of chaos: rafts for all the exiles who travelled away from the arid and warlike lands of late-modern capitalism.

Interview with Franco Berardi (Bifo)

by Giuseppina Mecchia
Bologna, 11 July 2005

Giuseppina Mecchia: *Why do you think that your 2001 book,* Félix, *is the first monograph devoted to Félix Guattari? Why was it written by an Italian intellectual and political activist? And when did you start writing it?*

Franco Berardi (Bifo): Well, let's start from the end. I thought about writing something on Félix exactly at the end of August 1992, just a few days after his death, while we were going to the Père Lachaise cemetery for the burial. I was walking with a friend of ours, Xavier Delcourt, and I told him that we had to start working on telling people about Félix. And I started right away and worked on it for a very long time. Actually, at a certain point, I talked to Paolo Fabbri, who at the time was teaching in Bologna, about the need to write something on Guattari. So, there isn't a specific time when I worked on the book, I kept writing for almost ten years. I started in 1992 and then I just kept writing and rewriting. Because, after all, it is not a theoretical book or an academic book: it is really a sort of narrative.

Then, why isn't there a body of work on Guattari? This is a good question, and there are several answers for it: first of all, it is true – even if we might regret it – that Guattari lives in the shadow of the couple Deleuze and Guattari. This is why there are many books on Deleuze that are actually books on Guattari. There is this flattening out of the couple Deleuze and Guattari on the part of the academic community, which has accepted Deleuze in its midst while Guattari is this embarrassing fellow traveller about whom much less is known.

But in fact, if you think about *Anti-Oedipus, A Thousand Plateaus* or *What is Philosophy?*, that is, the books they wrote together, you realize immediately that Guattari's contribution was truly decisive, even from a purely philosophical point of view. Surely from a linguistic point of view, in his way of handling concepts, but even from a philosophical perspective Guattari brings in an activism, a kind of materialism, a sense of the pragmatic aspect of concepts that, in a way, Deleuze didn't possess before ... And this is why I believe it is worth emphasizing their specificity.

GM : *But why an Italian book? Why you, and not a French intellectual? Why not a French friend of Guattari?*

Bifo: Because Guattari's friends were not philosophers. Well, he also had friends who were philosophers, but mostly his friends were artists, different kinds of druggies, crazies, and finally among his many, many friends he has been remembered in other ways. Jean-Jacques Lebel wrote plays in honour of Guattari, then there is a work of art which is a heart in a garbage bin, a machine full of plants. Gérard Fromanger, the painter who lives in Siena – one of the greatest contemporary French painters – created many works thinking about Guattari ... After his death, there has been a flourishing of these kinds of works. One also should remember that Guattari lived – and kept living even after his death – more as the guru of a community of oddballs than as a representative of academia. He's is not a *maître-à-penser* [master thinker], but rather, in a certain sense, a *maître-à-vivre* [life master].

But I also think that one could give another kind of answer. In the French publishing world, in the French cultural world, which in the last twenty years has suffered from a great crisis of cynicism, Guattari is seen as an interloper of sorts, as a madman who penetrated the philosophical field preaching deviant and heretical principles. We should state things clearly and just say that since the so-called 'New Philosophy', in France we have seen the growth of a deep political conformity which has spread to the vast majority of academia and the publishing world. The French publishing industry is extraordinarily conformist, certainly more than the Anglo-American one. A French publisher I spoke with just a few years ago told me that Deleuze and Guattari are certainly present in the French bookstores, but as writers belonging to a past era, not to the current one. Whereas anyone who has read Deleuze and Guattari knows very well that there are no authors that are more contemporary, or actually more 'future', than those two. In France, the idea is that quality resides in Bernard Henry-Lévy, which means the complete banalization of journalism, the abandonment of conceptual thought and, most of all, of any non-conformist way of thinking. These seem to be the best reasons to understand why the French philosophers don't talk about Guattari.

GM: *OK, and in Italy then? How did you find a publisher in Italy for this book?*

Bifo: Look, I never really tried to find a publisher; the publisher found me and the book. I wasn't looking for a publisher because I couldn't finish the book and so I was not ready to hand it over. As I told you, I was working on it without thinking that it would have to become public. It was the elaboration of an experience, of an apprenticeship, and this is why I never thought about looking for a publisher. Then I met Luca Sossella at the very moment that he had decided to get into publishing. Luca Sossella is a character from another era, an extraordinary aristocrat who had decided to do publicity for ten years in order to earn a lot of money, and then to invest it entirely in his own concept of poetic publishing. In fact, he is an editor who now is publishing poetry, with great success, I have to say, in the domain of poetry. He published something by Gassman, he published one of my ideas, and currently some things by Balestrini. So he is very devoted to poetry, and when he read *Félix*, he took it as a book of poetry, which made me very happy.

GM: *And in fact it is a very lyrical book.*

Bifo: Well, that was my intention, and this is how Luca read it and he published it with this great poetic enthusiasm that he has. Unfortunately, the book came out in a disastrous period, in May 2001, the very moment when the Right won the Italian elections. This is the period in which we still live, a moment of absolute dissipation in terms of intelligence and thought. I remember that we were sitting right here with Luca and the first copy of *Felix,* and we learned from the radio that Berlusconi had won. And here, I believe, is another reason why in Italy the book has had some circulation: that is, I believe that while the book sold out quickly, it has remained kind of a cloistered phenomenon, that is for friends and friends of friends. It did not become a public event.

GM: *On this point, I have another question: you say that the book sold rather well, but do you know if Guattari's thought or your book on Guattari is taught in Italy, in communication, philosophy or history programmes, in courses about French thought, or maybe in French departments?*

Bifo: I know enough about it to say that in the last few years, there has been a lot of attention given to Deleuze, some to Deleuze and Guattari, but it does not seem that anyone has devoted courses solely to Guattari. So now that you mention it to me it is the time to begin doing so!

GM: *You already told me that the situation in French publishing is one of excessive conformism. However, were you ever contacted by any French presses to have the French translation of your book?*

Bifo: No, and I didn't expect to either. Not this book. To go back to the Italian situation, there is something else that is worth saying: in France, Guattari, and even Deleuze in his final years, are regarded as a thing of the past, as something from which to be liberated, a residue from the Seventies which irritates those who would like to forget their past, who in fact have erased their past. In Italy, instead, Guattari is considered the friend of the terrorists. He's still the one who helped me first and Antonio Negri later. In 1977, the friendship between Guattari and me was considered the collaboration between two international terrorists ... And this image of Guattari as the friend of terrorists is what was created by the Italian press. Then, with the presence of Antonio Negri in the 1980s, he was done for. So, if today you mention Guattari to an Italian journalist, but also to a university professor, they will answer, 'Sure, Antonio's friend', and Antonio's name is actually an insult. This is very important in order to understand the Italian reception of Guattari.

GM: *To move on from publishing to some other questions, I found the book to be quite beautiful precisely because it combines a narrative with concepts, and I wondered if you had followed any models, philosophical or literary, in writing your book. That is, how did you find its tone, halfway between an essay and a story? Did you have any texts in mind, any literary or philosophical models?*

Bifo: Well, on the philosophical level and somewhat stylistically, I refer to the introduction of *What Is Philosophy?,* Deleuze and Guattari's final book, and to their discussion of friendship, which is sort of why I decided to write it in this way. And I would also say that I tried to return a bit to an idea of philosophy as an existential adventure that, even if from a linguistic and narrative perspective it might seem incongruous, I originally found in William Burroughs's characters, which were very well loved by Deleuze and Guattari, with their idea of the incarnation of the concept, the concept of existence, of suffering, of pleasure and so forth ... And then the literary references are so numerous that it is too hard to synthetize them.

GM: *So, the existentialization of philosophy and of the concept in William Burroughs, sure. But what do you think was the role at the end of the 1970s and then in the 1980s of Guattari's Italian contacts in the development of his thought? What do you think about his collaboration with Antonio Negri? I don't know if you you know anything about his work with Michael Hardt, that I think happened starting with his interest for Italy, among other things. What were the activities through which relations between Italy and France developed: psychiatric meetings, friendships, political commitments?*

Bifo: Well, Guattari came to Italy in the 1970s, in 1974 to be precise, to some meetings in Milan that Verdiglione organized,[1] and first of all in Basaglia's domain.[2] But there is a book which was the first thing that I saw with Guattari in it, from 1973, entitled *Psychoanalysis and Politics*, edited by Verdiglione.[3] These were the years when Verdiglione presented himself on the Milan scene as a cultural organizer, and so he organized this meeting in which Guattari participated. I believe its title was 'Signifying Semiotics'... Anyhow, at the beginning, he was following the Italian psychiatric circuit, in which Basaglia was the great reference

But then, in March 1977, the news arrived – and this has been told to me many times – that while Bologna exploded with the Autonomia movement, our reference point, and the reference point for organizing the Amsterdam Collective, our other collectives and even Radio Alice, was the fact that people had read *Anti-Oedipus*. Or anyway they had heard about it, as it was a rather well-distributed reference text. I read *Anti-Oedipus* in 1975, and I remember that the first meetings about Radio Alice already included references to *Anti-Oedipus*, when in reality, they had all read another book, by Guattari alone, that was published in Italian by Bertani in 1974 entitled *Una Tomba per Edipo*, and in French, *Un Tombeau pour Oedipe* [*A Grave for Oedipus*], with the subtitle *Psychyoanalysis and Transversality*. So let's say that they knew that in Bologna there were protesters who instead of crying Long Live Marx, or Mao, were shouting Long Live Deleuze, Down With Oedipus, Long Live Deleuze and Guattari.

I met them when I went to Paris in June 1977 following an arrest warrant, and immediately they told me: 'You Italians are greatly discussed in our milieu, since it is well known that Bologna is a place where *Anti-Oedipus* has had an impact.' So they were already paying attention to us, and then in the summer of 1977, there were a hundred or so persons imprisoned in Italy for different reasons. This is how Guattari undertook an initiative that resulted in the September meeting.[4] In July 1977, we wrote together the call that was distributed and circulated in Italy and elsewhere, and so his attention to the confrontations in Italy became increasingly a political awareness. In September, when he came to Bologna, the newspapers kind of presented him as someone challenging Italian powers, such as Bologna's mayor. So his attention to Italy became increasingly politically motivated, and this continued into the 1980s when after April 7, 1979, and the criminalization of the intellectuals of Autonomia, Guattari became – need I mention it? – a primary reference for anyone trying to escape Italy. I recall that the apartment on Rue de Condé where Félix lived, where I arrived in 1977, was that year a reference point for a sort of Italian-French *Soccorso Rosso* [Red Aid], and also for many Germans. Then came the attention for the communication issue: Radio Alice in 1977 had constituted a reference for discussion, but it was still more a political reference than a communication reference. But at the start of the 1980s, Guattari committed himself to the Radio Tomate event in Paris[5] that returned to many of the themes and practices of the Italian free radios. In the following years, there was the Minitel and then growing attention to the media issue.

So Guattari came to Italy in the 1980s many times. He had a very strong relationship with [the actress] Laura Betti and also with Tonino Guerra, Pasolini's and Fellini's scenarist. So he had these kinds of contacts that brought him often to Italy. Then at a certain point, I recall in 1985, he came to a conference on schizoanalysis in a kind of social centre that we called Topia: Centre for the Ecology of Mind. This was a very lovely conference on these themes. So in the 1980s Félix was henceforth a public figure who represented in Italy this European dimension of Autonomist thought, really a reference point in the more difficult period of cultural retreat [*riflusso culturale*].

GM: *For the Italian Left, yes, the famous retreat. And do you recall how Michael Hardt arrived here, with Guattari, how they met?*

Bifo: No, I don't know anything about this. I know that Michael Hardt was with Negri at the university in Paris, and I could be wrong, but I think that Michael had met Toni [Negri] and then became acquainted with Guattari. And then Michael wrote his book on Deleuze.

GM: *Why do you think that today, in the international academic world but especially in the Anglo-Saxon countries, there is a great renewal of interest in the Italian political thought of the 1970s? Why, at the beginning of the twenty-first century, are we talking about post-workerist thought, beyond Guattari if you like?*

Bifo: Well, this is the most important question, and it has nothing to do with the past, of course. It is about our present. I have the impression that we would like to understand what Anglo-American political thought has been looking for in French post-structuralism and Italian 'workerism', from the 1990s onward. Well, there are several trivial answers to this: the great success of *Empire*...

GM: *Yes, but why did it have this success? That is the question, right?*

Bifo: Precisely, this answer is not an answer. All right, I think that the issue is quite different. What is the conceptual innovation common to French post-structuralism and Italian 'workerism'? In my opinion, it is the rethinking, if not the abandonment, of the issue of subjectivity itself. Suddenly – well, actually it took about thirty years for the process to come to completion – suddenly subjectivity ceases to be what it had been in the modern tradition, from Hegel onward. It no longer is the strongest, founding element of the social experience, and starts to be understood as the product of a chemical composition of sorts, containing social – and this is what happens in the Italian workerist tradition – psychic, desiring and affective elements – and this is what happens in French post-structural thought. In other words, the strong element common to these two fields of thought is the rethinking of subjectivity, which is now totally emancipated from the rigidities of structuralism but also, and this is a more subtle passage, from historicism. As Guattari used to say, there is no subject, there is a subjective action, or rather a process that creates subjectivity.

Now, why does this impress us today, in 2005? I think that this happens because today, in the digital era, in the anthropological, more than political, mutation that we are experiencing, especially in the Anglo-American world, we are finally realizing that there is no subject. On the social and psychological stage there is a process of perpetual composition and decomposition of elements, of mixing singularities, which gives birth to temporary subjectivations. Now, I think that this is an element coming to the surface precisely in the era of the technologization of the social relation, that is, when the technological, the 'machinic' as Guattari would say, is increasingly entering the constitution of the communicative relations and even our corporeal experience. It is at this point that American thought, which knows the machinic much better both in theory and practice, understands that we have to rethink the notion of subjectivity. You should also keep in mind that Hegel, who had an essential role in German and Italian thought, is quite a mystery in America. He's either unknown or misunderstood. That is, the concept of totality escapes American pragmatist thought, and this makes American thought much more 'virginal' than the Italian or the German one with respect to the rethinking of subjectivity. And then you have more contingent, political reasons, but I really think that there is a fundamental philosophical issue at stake: technological and anthropological mutations are forcing us to rethink subjectivity. The subject is no longer a historical or a natural *datum*, but the product of a psychological and social process. This is what the French post-structuralists had discovered ...

GM: *But this is also what the 'workerist' thinkers had discovered and what marks the experience of* Autonomia Operaia, *which saw politics much more in the field of immediacy than in the Marxist project, which is oriented toward the long term and is historicist in nature. It was a practice of 'let us do politics for today, not for tomorrow'.*

Bifo: Yes, and here there are some considerations to be undertaken about the strictly political timeliness of that theoretical and practical experience, mostly the one of the late Seventies – from 1977 on, let's say – when the historicist framework, typical of classical, dialectical communist ideology, 'We are fighting for the Revolution', is fading away ...

GM: *We are fighing for the society of the future ...*

Bifo: For the constitution of the totality of the *Aufhebung*, for the totality of overcoming. When this horizon starts disappearing because of contingent, political reasons and also for deep, philosophical ones, with the deterioration of the dialectical framework, then the encounter with American pragmatism becomes much easier. In fact, the experience of the nineteenth-century European working movements, which was German, Hegelian in nature, always had great difficulty in understanding the American workers' movement. The American movement was populated by strange, mad individualists; people like the Wobblies in the 1920s who wanted their own freedom and a decent salary, but who didn't frame this project in a palingenetic, dialectical manner, tending toward the realization of an ideal. And this is why, I think, a squatter in Detroit is very similar to a squatter in San Basilio in 1978.[6] They start from the immediacy of a need but also from a very sophisticated immediacy of desire, that are completely independent from the dialectical expectations that motivated, but also constrained, the European workers' movements in the 1900s. This is why, today, it is much easier to establish a dialogue between American and Italian traditions. But these traditions, today, are no longer known to us.

GM: *Well, this also happens because they are not recognized as such: you and I are talking of an 'autonomist tradition', but if you mention it to other people ...*

Bifo: Yes, but this is a very interesting question: what do we mean today by 'European thought'? Because this also means asking what it means to be European now that the most important grounding for European thought, which basically was Hegelian dialectics, has collapsed. Today, European thought is much more differentiated, much more pragmatic than the German thought that had such a decisive influence all over Europe. In the twentieth century, European thought meant dialectics. Now, it means something much more similar to Anglo-American experimental thought.

This is not an answer, because I can't answer that kind of question. But I am trying to open up a debate, a very important one, because we are not talking about cultural 'fads'.

GM: *Well, this is really important, because in the United States this is the way it has been presented: here is the new cultural trend, 'workerism', we have gone from Derrida to Italian 'workerist' theory!*

Bifo: Not at all! We are facing the possibility of conceiving a de-centred world, and a non-dialectical future. That is, we are trying to ground the future, and politics, on desire and on a psychological and social composition that no longer has any historicist, logical framework. This is why this is an important issue, and not simply a cultural fad.

GM: *I think so too. Now, I have been wondering about the question of the role of the Italian Communist Party (PCI) in the 1970s. I was thinking that* *often in Autonomist thought, the PCI is considered somewhat like the party of historical compromise, the party at the basis of the great betrayal of workers' struggles and of workerism. Have you ever thought of a positive role that the PCI might have played inside the so-called 'sections' in neighbourhoods, as somehow the site of an affective investment in the political? That is, I am thinking of my mother, for example, who almost every evening went to 'the section'. That is, I don't think she went there simply because one day the revolution would happen, etc. It was somehow an aggregative site that allowed people to talk, right? And do you think that the collapse of the Communist Party eliminated something from the Italian political landscape, or was it simply a well-deserved collapse – who cares and so much for the better?*

Bifo: Well, to begin, my father was Communist and went every day to the section that we had in the basement, and at age fourteen, I registered in the Italian Communist Youth Federation. At sixteen, I was the secretary and representative of the Communist High-School Students, so I know this quite well as reality, I know what you are saying and I share it. But we are describing in this way the daily, existential reality of the Communist Party. We have spoken a lot about the forms of life, but not about the ideology, or even really about the the political function that the PCI effectively performed in Italy. And so, to answer your question: certainly, you are quite right, the PCI played a more fundamental role than other forces might have played elsewhere, finally the role of community, a type of communitarian element that we cannot minimize. The fact is that this communitarian reality became contradictory. Because one thing is the Party of the Roman working-class neighbourhoods, and another in the sections of Bologna, which were very different from Rome. In the Bologna sections, where many people I know very well were members in the PCI for years and years, they joined the PCI with the precise goal of having a career, of becoming city officers, of being able to manage matters ...

GM: *The 'Craxian' side of the PCI in Bologna, or the so-called 'Red administration'.*[7]

Bifo: So when you tell me about this reality of the Party as a community, I share it completely, but really this insight does not fully describe the Party reality, or what were then the Parties. I mean that I think this could also be said about the USSR: if we could know well the reality of the Communist Party in the USSR, we would discover that it was entirely ...

GM: *... that there were positive and constitutive sides of it ... So to speak precisely about an extremely contemporary matter, the question of communication, of the media, you are famous for having been one of the organizers, perhaps* the *organizer, of Radio Alice in the 1970s. So you were very aware of the importance of communication in political discourse and the discourse of subjectivation. When you met Guattari, either in person or in his works, were you inspired to keep working in this direction? In the Telestreet project that you and others created at the start of 2000, over the last few years?*[8]

Bifo: Certainly, Guattari's thought, especially in his later years, in the 1980s, played an illuminating role for all sorts of questions, for example the principle of the Web, that is, what Guattari called at one point 'the post-mediatic'. I do recall very well a meeting we had at Guattari's house near La Borde, around 1982-83. He told me: 'Note well that we must speak about the "post-mediatic era",' to which I said, 'But you're crazy, how can that be, since we are just entering the period in which television will rule everything?' So he said: 'No, no, that is only on the surface; the truth is much deeper.' And I recall that moment when he tried to convince me without succeeding. However, I was much more obsessed with centralizing powers than with the post-mediatic potential.

Ten years later, the Internet concerned me greatly from a practical, professional and even intellectual perspective, and in the Internet experience, I understood what Guattari was trying to say ten years earlier. So from the theoretical perspective, Guattari clearly anticipated questions of post-media and of the proliferation of sites as the decisive element in the mediatic game that is the Internet paradigm. From the practical perspective, perhaps things are different in the sense that, in France, the radio movement was a bit behind, born with great attention paid to the experiences we had in Bologna and, more generally, in Italy.

And also on the level of data transmission, you know that Minitel was born in 1981: hardly was it created when President Mitterand was elected and decided to informatize all of France by giving out a million Minitel terminals. From a certain perspective, this was an act of great foresight, but paradoxically Minitel blocked France rather than moving it forward. This is a story that is really worth being known since well before the USA, France had entered the terrain of mass information, but conceptualized it completely as a French, a national problem. At some point, the engineer who had constructed the Minitel system, Thierry Gaudin, went to Mitterand, or one of his people, to propose to the political powers that they cease using French and instead use English, that is, to make Minitel an open system. Naturally, he was reprimanded, and if it was even discussed, it was a French matter. And two or three years later, the World Wide Web was launched, on an international scale and conception. So for a number of years, France stayed behind. Minitel slowed up development, a paradoxical story that speaks volumes about French cultural nationalism and its power.

GM: *Can you say something about Telestreet? How was this project born, how was it developed, and is it still active?*

Bifo: I have to say no. Telestreet was born as an effect of dispersion, that is, when Berlusconi controlled everything. I was working for the radio, had been working for RAI for six years, and did a report on the Internet that had some success. Two months after the power grab by Mediaset – and I am not going into details now – I was kicked out along with hundreds of other people in the Italian communication system. At this point we were beginning to reflect on what we could do in this situation, and one of us, who was always the technological brain from the time of Radio Alice, said: 'Look, if we position an antenna on this roof, I have access to a kind of program that can transform it from a receiving antenna into a transmitter. Why don't we spread it around and everyone can do his own show from his house?' What a great idea! And in fact it worked. We proposed this idea in summer 2002 in Bologna, in a neighborhood nearby, and in six months, Telestreet had around 80 transmitters in Italy. In December we held a meeting, a large group attended, many participants in the media republic, many much younger than us, and so on. In 2003, the thing really exploded, reaching 150 Telestreets, some starting also to work via satellite, then the police closed two or three, but then reopened them. That is, a very interesting dynamic was unleashed at this period.

Then external problems started to arise: the first one was that you couldn't do a half hour or an hour daily transmission without funds since it meant mobilizing four or five persons to work on editing, I don't know, paying for the truck, and then it also meant investing in some technical equipment, not expensive, but still necessary. And so at a certain point, it began to really get expensive. That is, 2003 was the high point, when the war started, and right around February 15, at the moment of the great anti-military explosion, the Telestreets reached their peak ... But then, they ran out of breath, energy, funds, and some people began to think that we had entered a phase in which the power war was such that the communication tools were no longer adequate to act on this terrain. In 2004, there was a final meeting in Senigallia, in the studio of a Telestreet closed by the police, etc., and a split in the Telestreet circuit occurred since a certain number of Telestreets, such as the one in which I was participating, proposed to seek out relations with local institutions, with the township of Bologna or different townships in Tuscany, that had shown themselves able to obtain financing. The other Telestreets, in Rome for example, more closely linked to Indimedia, rejected this institutional manoeuvre. And at that point I was allied to the first group, but today I realize that the second one was right because, well, I spent a year waiting for the available funding to appear, and in reality, the imbecility of this ruling class on the Italian Left is such that they had said 'Great idea', but never let it be realized. I would like to see when there will be a Centre-Left government, if they will be able to deal with the fact that Berlusconi will still dispose of six television networks, because how can we think of doing the same epuration of the RAI that he did when he came to power? No, I do not think it will be possible.[9]

GM: *Neither do I. They did not have this willpower.*

Bifo: No, and in any case, the power that Berlusconi accumulated in the communication field is not solely the power of organizing the television programmes, but also the power of advertising contracts, and the corruption of the leadership structure in the entire communication system. So, strengthening the Telestreet circuit would also serve, I think, the Centre-Left, but as usual, this was not understood, and lately the Telestreet circuit has been in the process of shutting down. But it may be redeveloped later and in any case, it was an important experiment, a liberatory one on the political and cultural levels, and also on the technological level since there were thousands of persons who learned how to do television, to raise an antenna, to do editing, and really learned the spirit of televisual communication.

GM: *Which does not necessarily need to be centralized since it can be taken in hand ...*

Bifo: Correct. However, Autonomedia is publishing a book dedicated to the Telestreet experience, the title of which should be *Mediodemocracy*. I read its proofs recently.

GM: *So let us return to Guattari. One of the questions that we have is about Guattari's language. That is, over twenty-five years his writing maintained a high level of complexity, with many neologisms, and he created temporary and flexible semiotic systems, within which the same word could change its meaning. How did you find your bearings in this complexity? When you wrote your book* Félix, *what was the process you followed? Did you just read all Guattari, and did you undertake a philological study? Or did you proceed like they described working on* A Thousand Plateaus, *that is, did you skip around and grab what you could?*

Bifo: OK, first, about the neologistic machine: this continuous invention that they theorized – and I say they because Deleuze is important in thinking about these questions, even if the invention of words was due more to Guattari than to Deleuze. The neologistic machine is not simply a stylistic oddity: it really is the idea according to which concepts are tools that allow us to cut up reality according to desire, need ... So inventing and constructing a word means elaborating something, means grasping tools in order to see things in a different way.

GM: *This is an enormous undertaking in reality.*

Bifo: Absolutely. It is true, the creation of concepts, of words that constitute concepts is also something that allows you to understand the ways that literature, philosophy and poetry fundamentally do the same work.

GM: *So how did you manage to engage with this discourse?*

Bifo: Well you asked me how I read them. I think I read lots of Deleuze and Guattari in the 1970s and the beginning of the 1980s. Then, when I started to write this work, my relationship with their texts became really impressionistic, here and there, a little like a tool box. While the neologistic activity, something that I certainly learned from Deleuze and Guattari – especially Guattari – this type of invention of words is something I really tried to shift into the domain of cyberculture. That is, to invent concepts capable of expressing the complexity of the relationship between technology and society. In the 1990s words began to proliferate, words that tried to conceptualize this new techno-social territory, and this is something that certainly came from Guattari's lessons.

GM: *And you also learned to some extent to write in this way.*

Bifo: I would really be pleased if I were able to do so.

GM: *Well, another question we had was about the collaboration between Deleuze and Guattari, Deleuze without Guattari, Guattari without Deleuze, but basically, your book is devoted to Guattari, to Félix.*

Bifo: Yes, and there's a long chapter on Deleuze.

GM: *Yes, but we were wondering to what extent Deleuze was important for you. That is, would you have written a book on Deleuze?*

Bifo: No! Because I am not the most suitable person to write a book on Deleuze. Let's say that some things in Deleuze are extremely important for me, and just to mention one for all, *Nietzsche and Philosophy*. This is a political book, philosophical but also political, of fundamental importance. Perhaps it is Deleuze's clearest book, but it is also a book from 1962, so while it was really early, it anticipated many themes that are now extremely timely on both the political and social planes. But basically I do not consider myself to be sufficiently a philosopher in order to address Deleuze in a direct way. While I seem to be just dirty enough from the disciplinary point of view, rather contaminated – I don't really know what I am – just confused enough in my disciplinary background to be able to approach this ...

GM: *Well, Guattari was also rather mixed from the disciplinary perspective.*

Bifo: Guattari was educated as a pharmacist, that is, he was someone trained in mixing substances. And when he spoke, often he sounded like a plumber, with all these things about flows, tubes, cutting, tightening. That is, Guattari was truly the philosopher who tried to make the linguistic, conceptual machine work in relation to the links to existence, to the social, and so forth. In this sense, I feel much closer to him. This is not only why Guattari greatly influenced my development, but I met him when I was 27, so I was already sufficiently trained to feel him as someone I understood perfectly.

GM: *OK, and beyond what we know about their way of collaborating, about what Deleuze has said, for example, speaking in the 'vous' form, things we know about, do you have any knowledge about the modalities of writing that they adopted while working together, any anecdotes about how they wrote together?*

Bifo: Well, from what I understood, Félix himself was very discreet about this, protective. The topic of his relationship with Deleuze was basically a rather secret matter. I met Deleuze at Guattari's place only one time because their meetings were not a public thing. I seem to have understood from Félix – because the question of collaborative writing is a question that interests me greatly, even recently, for a thousand reasons – that their procedure, let's call it, was like script-writing: they worked out the script and then worked on it separately, after which they would meet and work, not so much to correct but rather to integrate. This is a method I would call recombinant, rather in the sense of Raymond Queneau, that is, the idea according to which there is an automatism in the language process that is derived from understanding on another level. That is, there is no need to polish off, but rather a need to combine, I seems to me. I would like to know more about it.

GM: *So basically it's a beautiful gamble writing together, it seems – that is, it becomes a gamble.*

Bifo: I don't know if you know Wu-Ming, formerly known as Luther Blissett ... [10]

GM: *Well I do not know them personally but I have heard about them.*

Bifo: I know them well, and I am curious and have tried to understand how they do it. We are great friends, and I have asked many times how do they do it? And this is how they proceed, that is, with a script, each of them writing it all, and then they see what works well. So this is a method that is not one of correction, but of combination.

GM: *To move a bit toward the present. From a conceptual perspective on things that are useful now, you have this concept of the 'semiochemical' that I found interesting. You make it clear that this is the way in which Deleuze and Guattari interpret the concept of the sign. That is, the sign is not only a convention, but also has a corporeal existence, a real effect on bodies, and you speak also about viruses. So how do you find this concept of the semio-chemical – that is, a concept of direct production, from the point of view of capital – now that we are in the era of cognitive capitalism, or the immateriality of work, when you no longer produce machines but advertisements, and the ad is not as a pure sign, but something that also changes bodies? Do you find that this semiochemical concept might be combined, or at least do you* *see it as resonating with Maurizio Lazzarato's thoughts on the immaterial, on the cognitive phase of capitalism?*

Bifo: OK, first of all, the concept itself: in *A Thousand Plateaus*, there is a chapter devoted to the double articulation, in which they talk about Hjelmslev, and the double articulation is precisely this. Language is this: on the one hand, it has a conventional, intangible character, but on the other hand, words are also gestural, factual, affective, and let's say that this motivates all the things you say about publicity as a circuit of material production. Then there is this strand opened by Lazzarato, but I can trace it back to French thought if we think about Lyotard. So I would respond to you: of course they connect. Except that I would prefer substituting the word 'immaterial' with other words, because really if we believe in the fact that language and signs possess their own materiality, if we believe in the fact that there is an affective, productive dimension – productive on the economic level, on the affective level – in the dimension of semiotics, then why employ the notion of 'immateriality'? I understand the problem, but then we use a word that allows us to situate the question: we are speaking about the cognitive. The cognitive dimension, cognitive work enters into determining the symbolic and the immaterial dimension of the social, for which I would say they are on the same level of discourse that Lazzarato proposed, except that I find that the word 'immaterial' seems to lead us off track. I understand, I do remember how it came about since at some point, people started talking about the immaterial. And they did so to say that an operation of a symbolic kind takes place – that's fine. But what we want to grasp and lies at the very heart of our interest is neither the mechanical nor the physical being, but the material being of what is symbolic! But about the word 'immaterial', we have already understood each other. Now if Lazzarato were here, he would say, 'OK, fine, but we can use "immaterial" to account for a displacement of production from the physical to the symbolic dimension'.

GM: *Yes and you know, in his latest work he becomes, if you will, a little less, not optimistic, but a little less triumphalist about the potentialities of immaterial work. Because at a certain point, it seemed that this immaterial work was a moment of liberation, whereas now he says that there are also forms of exploitation in immaterial labour: there is immaterial labour also for those who serve as elder care providers. Because this is also immaterial labour, not only computers. Immaterial work is also that of a receptionist at a hotel, who says hello, but it's an underpaid form of immaterial work.*

Because at a certain point, it seemed that immaterial work was exempt from exploitation.

Bifo: Yes, but on this point work has been done for twenty years now, and so in some ways, the concept of 'immaterial' has been left behind to make room for a post-industrial direction of the reflection about work. We have moved on, even if it is true that Lazzarato continues to use this experession.

GM: *Another thing that interests me is when at the start of your book, in reality you are talking about Guattari's commitment to the ecological movement of the 1980s, and it seemed to you to be an act that masked his depression, a nearly desperate act that arose from what you call a certain end of political will. Why? What was missing in the Greens' commitment then to be recognized as fully political?*

Bifo: But I have nothing against the Greens!

GM: *But you do say there was a way ... that basically Guattari did not believe in it himself, and that you yourself did not want to commit to this.*

Bifo: Surely in this, there was also a bit of my own idleness, my own laziness. But I sensed and was convinced that in the final decade of his life, Guattari had at several points undertaken a political commitment in which he did not deeply believe, that is, seeming to him to be his duty to 'hold on', that he needed to get past this rather difficult, regressive period, etc. And I perceived a kind of exhaustion in his will to maintain a position. So in this phase of the Guattarian itinerary, what seemed to me to be missing – as it seems missing in much of Toni Negri's thought – is a reflection about depression. While one would need to enter more fully into this concept, depression basically is a disinvestment of libidinal energies in facing the future, in facing the world. Naturally it's a question of a pathology, but not only that. Or rather, in short, the pathology is not something to be undervalued.

GM: *Or repressed.*

Bifo: In depression, there is a cognitive, philosophical and political content that, I feel, was not fully developed by Guattari.

GM: *Do you think that his meeting with Negri might have prevented that from happening?*

Bifo: Certainly, certainly! That is, the relationship between Negri and Guattari is the encounter of two old militants who had seen so much, and who at that moment, that is, in the 1980s, were living through a moment of defeat, of difficulty, but as 'real men'. And there one finds a 'macho' element ...

GM: *In Negri, certainly, but in Guattari, I had never felt that.*

Bifo: It assumes another form: see, perhaps if Guattari had not met Negri, he would have been able to take stock of his aging, he would have faced his feminine side and his depression. But instead, this encounter became for him an element that reinforced his will toward maintaining an historical presence. It is the Promethean, heroic element that is very strong in workerist thinking. Because we have to admit it, it's a 'true men' mode of thinking. I think that all this became, I would not say a kind of impoverishment, but definitely there was something missing. Guattari, among other things, entered into a phase of personal depression, of a very profound personal crisis.

GM: *But he was taking medicine, right? He wanted to feel better, that is, he did not want to come to terms from the analytical perspective, you see, not to go into analysis, so he was taking medication to feel happier.*

Bifo: And he did well to take medication, but perhaps beyond that, I think he should have also considered that perhaps...

GM: *That there are reasons to feel depressed ...*

Bifo: Yes, and on this, one would need to undertake a long discussion, to consider relations with women, with men, and basically politics as an essentialy masculine enterprise, beyond words, beyond intellectual discoveries that perhaps remain purely intellectual. But I don't mean that Guattari was macho, because in fact, it's not true. But in the political and also the philosophical dimension of the last part of his life, there was the prevalence of the sense of historical presence, of historical duty, of the dialectic, which made it impossible for him to grasp the cognitive element that is present within depression. Depression teaches you something, and you shouldn't forget what you've learned in depression. And then, nothing, a glimmer of awareness came forth in the introduction of *What Is Philosophy?*, when Deleuze and Guattari speak about old age – that old age allows us to begin to understand. Guattari was close to 60 years old.

GM: *In this, I hear a lot of Deleuze, who was older and certainly having greater illnesses.*

Bifo: There's also Guattari, who in his final years lived through a very painful sentimental experience.

GM: *Right, with Joséphine, something I knew nothing about.*

Bifo: Well, she's his companion with whom he lived during his final years. She was a lot younger than him. Well, Joséphine also came from an experience of drug addiction, so in some ways, she also represented for Guattari a psychotherapeutic challenge, not only a sentimental and sexual one. And so that seemed to me an interior event that he chose not to elaborate except in this moment in the introduction of *What Is Philosophy?* It's an embarrassing topic to take on because really, as I know well, the Guattarian network did not want to talk about her. She was perceived as a foreign body: she was quite young, she became 'the boss's girlfriend', and so she was very isolated from Guattari's friends. And then Guattari became isolated from them as well, and so a break occurred: those who remained with Joséphine, very few, but among whom I count myself, and many who considered her an idiot who should not be there ... And so even in this, the relationship with femininity and with aging, with depression, became a crisis factor within the political community.

So I have spoken about all of this briefly in the introduction of *Félix*, for it might not be very interesting from the existential perspective, but it interests me greatly from the philosophical and, I would say, also from the political viewpoint. For historicist thinking and militant practice refuse to consider depression as a cognitive element, and this is a limit, one that today, for example, prevents us from being lucid about understanding the collapse of modern hope. We should understand that the collapse of modern hope is certainly a disaster, but that it also contains elements that we should succeed in understanding, granted that our humanistic, socialist, illuminist, communist values no longer have any place. And this is depression, when you realize that your desire no longer has any place in the real. This is the deep core of depression. If you insist in not wanting to see this fact, you end up continuing to use tools that prevent you from acting.

GM: *Yes, even Antonio Negri, who claimed to have moved past these things, continues to use the concept of revolution. While in his later work, like* Kairos *and* Alma Venus,[11] *where he actually mentions St Francis of Assisi, he talks about the need to discuss the end of certain things, and he also deploys some concepts that are no longer relevant, basically the concept of a revolution that, honestly, I don't know who would do, and where.*

Bifo: And at this moment optimism of the will seems to me a kind of hysterical reflex.

GM: *This did not seem so much a characteristic of Guattari. It's true that perhaps it was revealed in his meeting with Negri because later, in the 1980s, he wrote some beautiful texts about the present, like* The Three Ecologies, *that are not triumphalist, but are rather darker, even depressive; that is, the end of* The Three Ecologies *is rather sad.*

Bifo: It's not a rebuke, but simply a consideration of the fact that Guattari's experience brings us to the limit of this reality of depression, it hints toward it, because it is shown in *What Is Philosophy?*, but he doesn't really get into it to valorize it.

GM: *Back to the question about the Greens and the 'no global' movement, what do you think about it? That is, it seems to me that the anti-global movement – with its questioning of the International Monetary Fund, or of so-called 'politics of development' of the planet – it seems to me that these movements have emerged to some extent from the inspiration of the Greens in the 1980s, that is, from the shift of focus from national politics to the social and political dimension of the Earth. Does it seem to you that this is a possible political terrain, or do you have doubts about this kind of engagement, does it satisfy you or not?*

Bifo: Well, on the conceptual level, I agree completely with you. The Greens' kind of thinking, ecological and environmental thinking, certainly has allowed a necessary displacement. But the so-called 'anti-global' movement, from Seattle onward, my impression is that this did not succeed in becoming a political proposal. It remained rather a discourse of critique in opposition to capitalist globalization, but up to now we are struggling to grasp its propositional, positive content.

Except for Latouche's hypothesis, which is the concept of postdevelopment.[12] The paradigm of postdevelopment is the discovery that modern history is based on prejudice – conceptual, political, psychic, mental – according to which we must expand the mass of consumers and the mass of production.

GM: *So in fact, you are basically saying that the idea of postdevelopment is a questioning of what has been an economic vision of politics?*

Bifo: That is, the domination of economy over society.

GM: *And what do you think? That is, do you think it is possible to go backward – or maybe forward – with respect to the last three or four centuries?*

Bifo: I don't know if it's possible, but I am certain that it is indispensable.

GM *Because Guattari spoke about this in* The Three Ecologies, *with great foresight!*

Bifo: Yes, when he allied himself with the Greens in the 1980s, he was absolutely right in doing so, on the philosophical level. It was coherent with his history and his general engagement. I never questioned this, and I fully shared his philosophical predilection for the Greens. However, what seemed to be a bit useless – and in fact, it amounted to nothing, I must say – was his militant generosity for which, at the time he was elaborating this idea, he began to do neighbourhood work with a bunch of cretins ... In short, the French Greens were not always the most brilliant ... He became a candidate for the elections, a candidacy that no one supported. So these seemed to me things that he didn't need to be doing, it was pretty much a waste of time. Overall I was embarrassed to tell him anything like that because at the time, I was twenty years younger than him, so how could I bring myself to say that and to give advice to Guattari? He might very well send me packing ... However, I recall that at that time, in 1983 or 1984, he insisted that I should present myself as a candidate, and he encouraged me to do this, either with the Radical Party or with the Green Party. Then he met Pannella, and Pannella said that I should become a candidate for the mayor position in Bologna![13] And in fact, one day Pannella came to Bologna to propose my candidacy for the adminstrative elections, and I told him 'Forget about it! No way!'. In the 1980s, I was in a

Bifo: Think about the things that could serve the entire movement to absorb the conceptual energy that comes from the Greens. It is useless to think that we are militants: militancy per se will not save us.

GM: *Yes, I understand perfectly. That is, the Green practice is much more efficient and important than parliamentary representation might be ... That is very interesting. Perhaps this is even a remnant, if you will, of your engagement in the Autonomist movement.*

Bifo: Certainly.

GM: *Practice is therefore much more important than parliamentary representation. And I hope so, that is, given that representative democracy in America is truly laughable, one must gather one's strengths and say 'But after all, politics does not exhaust itself in this'.*

A bit more about the present now, that is, the present that goes beyond the writing of Félix. *In your book, you talk about Guattari's reflection, in my view very intelligent, inspired by Sartre, on groups of political activism or creation in which there might be a difference between the subjugated group and the subject or creator group. For if a group is not subjugated, singularities are expressed, but if the group is subjugated, singularities get reterritorialized, there is even a fascist drift with collective activities. How does one refute this tendency, that is, this group tendency to become subjugated? When does it happen that singularities do not get caught in the fascist or neo-fascist derivation? When one speaks of the old terrorism of the Red Brigades, or when one speaks of contemporary terrorism, and we get beyond the manipulations, the secret services, etc., there must be something there that allows groups to become subjugated. How can one resist this tendency?*

Bifo: You cited Sartre who spoke about 'groups in fusion' and of the 'practico-inert' in order to say that the group in fusion, which resembles the subject group that Guattari discussed, is the dimension of the collectivity in which the collective is a choice, or rather a desire, a dimension in which desire is expressed. While the practico-inert is the dimension in which group activity is suddenly a function of external requirements and automatisms. The pressures of economic production, for instance, or, and here I am getting to your question, the fear of conflicts with the outside. Today I would say that the decisive element in the question you have posed is that of identity, of the identitarian obsession. We can say that the singularities create the collectivity in a desiring manner until this collectivity essentially represents an opening towards the other, the pleasure of discovering the other as a potentiality of your own being, as consciousness ... Up to this point, all is well. However, then there is a phase, not necessarily a chronological question, but a different modality of what Deleuze and Guattari call 'becoming other', in which you find in the other a reinforcement of your own identity. You are equal to me, and therefore we can together set ourselves against whomever, in our conflicts, is the other. Or you confirm my identity that I know, as a worker, as a male ...

GM: *Or female ...*

Bifo: Of course, as well. Sometimes the confirmation of identity is a confirmation of one's own dependence, of one's own weakness, of a condition of subjection that is deeply inscribed in my being that I am no longer able to recognize myself in anything but my own condition of dependence. So here, I would say that the passage from the group in fusion to the practico-inert has identitarian forces as the decisive element. Now is there a cure for a passage of this sort? I would say we have to be cured from the identitarian need, which is deeply inscribed in masculine, patriarchal history. That is, I know very well that there is a need for identity also for women, and I see it, I understand it and feel it, but it seems to me that the social sanction on identity is essentially a masculine phenomenon; it's from the male that the necessity to stabilize identity as a positive element arises, as an element that in a certain way constitutes an acquisition that it's frightening to lose. On the female side, identity is simply a lifeline, a way not to lose everything, while the female element is essentially dispersive, opening itself to the other. I don't mean here, I am not speaking about men and women, but rather about the male and female as cultural qualities, as modes of being, also as a way of seeing the other. So here, I have the impression that the problem of keeping open the ability to see the other is the same as the ability to reduce the identiarian anxiety. Reducing this anxiety is a problem perhaps that does not belong essentially to politics, but belongs rather, not to psychotherapy, but to schizoanalysis! What is schizoanalysis? It's the capacity to use the tools from politics, psychotherapy, art, language, everything, in order to succeed in keeping open the capacity to be other, to become other, to take pleasure in the other. Now that thing, that passage is one that is made possible through an action of an essentially therapeutic character. But it is also the deep core of politics ... and it is a cure for the increasing rigidity of Western politics. And not only Western, naturally.

GM: *Yes, of course, since for example, in America, this identitarian desire that you discuss, this identitarian necessity, gets stronger every day in my view and to my great embarrassment and fear, finally in the sense that there is this profound sense of belonging to a church, to a race; people close themselves into these groups where they demand that the other be like us. And I wonder if schizoanalysis as Guattari developed it ... that is, if we are not now in another reactionary phase ...*

Bifo: Well I myself think so. That is, it's clear that what was happening in America in the last few years ...

GM: *But also in Europe ...*

Bifo: It's clear, the reaction to the events in London [the bombings of July 2005] is impressive in this sense: it's different from Madrid ... Madrid had the effect of increasing understanding. The train station bombing in Madrid allowed the Spanish to understand that if we wage war, we are entering into a very dangerous and brutal situation.

GM: *Yes, and in fact, the Spanish understood that.*

Bifo: And this helped all Europeans to open their minds to an absolutely obvious thing, absolutely elementary. However, the London events caused us to undersand that Islamic terrorism can become a nearly daily fact of life. Even in the technical modalities of this operation: this was not an operation of great complexity ...

GM: *Yes, it was not like the Twin Towers, where at least you had to coordinate the plane schedules ...*

Bifo: Yes, in London there was nothing to coordinate! You just have to place low-yield bombs and you are still going to kill people, no matter how. So there is this perception of the fact that it's something that can be multiplied infinitely. Now, faced however with this, instead of reducing tension, instead of saying 'We have to avoid a growing identitarion counter-position, because this is causing our deaths, not just theirs', instead you see what the reaction is? The reaction is to say, 'OK, systematically they want to kill us, so systematically we are going to kill them'. So now ...

GM: *This has been the Bush policy, among others. Except that Bush, I think, lied knowing that he was lying, when he said 'We are going to war because they are killing us and want to destroy us, so we will destroy them before they do'. I think he lied knowing he was lying.*

Bifo: I really can't say. On this, I don't know if Bush in particular ...

GM: *Well, people like Cheney, for example ... but not Americans in general ...*

Bifo: I am not able to judge as to whether there was fanaticism in the strict sense, that is, about the possibility of obtaining something, of overvaluing one's own power, or whether they were just in bad faith. It is clear that the choice made by the Bush administration drove the entire West into hell. But who knows if there was from the very start an awareness that things would have gone like this, or if perhaps they are such imbeciles to think that you make a lovely war, killing five thousand people, and at that point they are all bowing in front of you asking for forgiveness!

GM: *Yes, I wonder if finally the American public is not a bit smarter. Then there are people who nonetheless say that the plan to invade Iraq was already there, with the intention of the US to get their hands on the oil reserves and that the Twin Towers attack was really all they needed to implement it. Anyway, now we are waging holy war, we are in this identarian necessity for counter-position, we are giving people the chance to feel that we are better than them. However that might be, I think that schizoanalysis is the only way to get beyond this ...*

Bifo: At least to elaborate the problem!

GM: *I believe so quite strongly. So, to reach a conclusion, you said at the start of the book – and perhaps you have already partially responded to this – that when Félix spoke to you about committing yourself politically to the Greens or the Radicals, you said that already since 1977 you had stopped considering yourself an activist and a political miltiant. So what about now? Do you consider yourself a political militant?*

Bifo: Let's say, from a pragmatic, factual, biographical point of view, certainly around the period of the Seattle events, from 1999 to 2003, I lived through a phase of extremely intense militant reactivation. Today

I would say that I am recently again in a phase of mobilizing militant energies, if I may express it like that. But beyond the fact that each person can find his or her energies in certain moments and less so at other times, or that each person thinks he or she can be useful at some moments more than in others, beyond this, I would say that already during 1977 I began to be convinced about something that I then learned at the heart of *Anti-Oedipus*: the fact that our task is not to put our will to work. Our task is to create modes of catalysis, translation and transmission of a socially dispersed desire. Thus, it is not so important that you devote four hours a day to your activity of convincing people of this or that. It is more imporant that in your existence, your work, the actions you accomplish, etc., you succeed in doing something, as Wu-Ming would say, of a mytho-poetic character, that is, succeed in condensing social desire in mytho-poetic form. Mytho-poetic would be precisely in the sense of producing a shared narration, that is like 'Look, this guy is doing something that makes him happy, that makes him content, makes his life fun: why don't we also do what he is doing?' This is what we should succeed in doing. It's more of an exemplary role, almost a monastic one if you will, precisely the idea that it is not about being militants, about carrying the truth to those who never received it: no, it is about being happy, about communicating happiness: this is the true duty of a militant. We don't always succeed. It is not enough to want it in order to succeed.

GM: *We aren't always happy either ...*

Bifo: Certainly, but the real political problem for me is this: to succeed in communicating mytho-poetically, that is, through a narration that might possibly be shared, a possibility of happiness. Happiness then is a word that nobody knows what it means, a possibility of harmony with the flowing of time, with one's own being in relations, etcetera.

GM: *This then was finally what Félix Guattari was doing. He wanted to be one with the present. To do so, he never ever gave in to nostalgia, did not lose himself in reflecting about the past. The present was for him the most important thing.*

Notes

[As Bifo provides no footnotes in *Félix*, all notes are provided by the translators; full references to works cited are in the Bibliography]

Preface

1. Giuseppina Mecchia would like to acknowledge and thank her colleagues in the French and Italian Department at the University of Pittsburgh for their continued support. Charles J. Stivale would like to acknowledge the support of the Humanities Center at Wayne State University for its support of this project.
2. Notable exceptions to this lack of engagement are by Gary Genosko, *Félix Guattari, An Aberrant Introduction* (2002) and *The Party Without Bosses. Lessons on Anti-Capitalism from Félix Guattari and Luís Inácio 'Lula' da Silva* (2003).
3. The recent interest in this movement has mostly sprung from the success of *Empire* (2000), the book authored by Michael Hardt and Antonio Negri. Since then, the works of Tronti, Negri, Panzieri and to a minor extent of Bifo himself have been the object of a sustained critical and political debate.
4. Capital, according to the theorists of *Potere Operaio*, was not the true leading force of economic and political development: labour came first, and if in fact labour came to be subjugated, this would still not cancel its ontological primacy in the production of goods, structures, and even of cultural practices. See for instance Antonio Negri in the essay 'The Constitution of Time': 'Behind the category of relative surplus-value hide the movements of productive co-operation that – originally (it should be forcefully underlined) – presents itself as the refusal of capitalist command over production and as the attempt, always frustrated but not less real, of constructing an autonomous time' (*Time for Revolution*, 73).
5. It is worth remembering that in 1970 the University of Bologna inaugurated the DAMS (Dipartimento di Arte, Musica e Spettacolo), the first and only academic programme in Italy where contemporary forms of communication were being studied together with more traditional performance arts. DAMS attracted young performers and aspiring social and cultural critics from all over Italy, and figures such as Umberto Eco delivered memorable and crowded seminar lectures.
6. A detailed assessment of Guattari's involvement with these post-mediatic initiatives is to be found in the article by Bernard Prince and François Videcoq (2006) entitled 'Félix Guattari et les agencements post-media'.

Chapter 1

1. Bifo refers to the Global Action Day that occurred on 30 November 1999, a broad range of demonstrations organized world-wide in protest against the

World Trade Organization meeting scheduled that day in Seattle, Washington. For a summary of events, see the Global Action Day reports at www.nadir.org/nadir/ initiativ/agp/free/seattle/n30/index.htm .

2. See the Collective A/traverso's (1980) statement about free radios, 'Radio Alice – Free Radio'.
3. The unrest to which Bifo refers corresponds, in general, to student demonstrations against State oppression organized by the Autonomia Movement in Rome, Palermo, Naples, Florence, Torino and Bologna in early 1977, especially in March, following which Bifo fled to avoid arrest. His account of these events appears in Berardi (Bifo), the *Semiotext(e)* essay 'Anatomy of Autonomy' (1980). See the website Affinity Project (1980) for summaries of the Autonomist Movement and its leaders (including Bifo).
4. In 'Anatomy of Autonomy' Bifo describes this September Convention as 'the great opportunity – missed, however – for the Movement to overcome its purely negative, destructive connotations, and formulate a programmatic position for the autonomous organization of a real society against the State ... Unfortunately, the Convention turned into a reunion against repression, and this greatly reduced the theoretical importance and the possibilities of this period ... The gathering concluded without producing any direction for the future, any new program, and without advancing the Movement' (59–60).
5. Guattari's account of this political repression is collected in the essays comprising *Les Années d'hiver, 1980–1985* (1986).

Chapter 2

1. In the original title for the chapter – *La depressione Félix* – the author plays with the literal meaning of Guattari's first name, *felix*, which in Latin means 'happy'.
2. See Chapter 1, footnote 6.
3. In English in the original.
4. The reference is to a retrospective exhibition of works by the Swiss sculptor Jean Tinguely at the Paris Beaubourg museum in 1988.
5. In French in the original. Once again, Bifo is alluding to the book *Les Années d'hiver* (1986).
6. Bifo alludes to the book by Philip K. Dick, *Do Androids Dream of Electric Sheep?,* where Dick defines kipple as 'unwanted or useless junk that tends to reproduce itself' (Dick 1968). It is a great threat that risks submerging the Earth.
7. This name refers to the *Tibetan Book of the Dead*. For a Deleuzian relation to the text, see Robert Glass (2001).

Chapter 3

1. Except for the final paragraph, this chapter serves as introduction to Guattari's 1977 book *Piano sul pianeta* (vii–xiv).
2. The reference is to the struggles of the Solidarity Movement in the 1980s against Soviet control.
3. Another, more concise text related to this one is Guattari's 1979 'Plan for the Planet'.

4. See also Guattari (with Eric Alliez) 'Capitalistic Systems, Structures and Processes'.

Chapter 4

1. Bifo's use of the term 'infospheric' refers to the vast information web developed by cyberculture. See also Bifo's essay on 'Schizo-Economy'.
2. On the UN-sponsored Conference on Environment and Development, see www.un.org/geninfo/bp/enviro.html .
3. This quote relates most directly to George's *The Debt Boomerang* and also her *A Fate Worse Than Debt,* especially Chapter 10, 155–68.

Chapter 5

1. While some may find this formulation disturbing, Bifo does not diminish the importance and severity of the AIDS epidemic. He emphasizes, rather, the 'name of AIDS' in the same fashion that Deleuze and Guattari employ schizophrenia as a semiotic process that has become generalized as a complex form of cultural production, parallel to the specific medical conditions associated with the particular name.
2. Guattari's writings on two of these artists can be located as follows: on Lebel, 'Jean-Jacques Lebel: Painter of Transversality', and 'Gérard Fromanger, la nuit, le jour'.
3. Guattari develops his ecological analysis in the *The Three Ecologies.*

Chapter 7

1. Bifo erroneously writes December 1996.
2. The reference within the citation is from the French edition of *Will to Power, La Volonté de puissance* (Nietzsche 1935,1937)
3. The internal citation is to Proust's reflection on the Narrator's collision with reminiscences, 'the fortuitous, inevitable way in which sensation had been encountered [which] governed the truth of the past that it resuscitated' (*Proust and Signs*, 96; Proust, *Time Regained*, 274).
4. We find, in fact, no specific reference to Deleuze evident in Guattari's text, as Bifo suggests.
5. This screen, says Deleuze, 'is like the infinitely refined machine that is the basis of Nature' (*The Fold*, 77).

Chapter 8

1. On biographical details for this period, see François Dosse (2007), 208–18. Guattari no doubt marched with the students and workers during the May '68 events; the extent of Deleuze's actual participation was more limited given both his ill health at the time, an also his location at the Université de Lyon, rather than in Paris.

2. This expression, '*adaequatio rei ad intellectum*', is one of two ways of understanding the traditional conception of truth as the correspondence between the thing and intellect. The other way is '*adaequatio intellectus ad rem*' (correspondence of intellect to the thing). See Fuenmayor (6–7) for concise elucidation of this distinction.
3. The English translation (1997 [1941]) of Marcuse's *Reason and Revolution* does not include the short text written by Marcuse in 1960, 'Una nota sulla dialettica' (A Note on the Dialectic) included in the Italian edition (Marcuse 1941, 43–52) and quoted by Bifo.

Chapter 9

1. An earlier version of sections of the rest of this chapter appear in Bifo's contribution to the collection *Il Secolo Deleuziano* (Vaccaro 1997).
2. 'Ah Pook Is Here' is a short story by Burroughs that was also recorded toward the end of his life and was used in the soundtrack of an animated film by Philip Hunt; see http://zed.cbc.ca/go?c=contentPage&CONTENT_ID=18034 (accessed 30 August 2007).
3. In fact, as we indicated in the Preface, the respective contributions to rhizomatic thought can be studied in Guattari's *The Anti-Oedipus Papers*.
4. In fact, *Bergsonism* came out with Presses universitaires de France in 1966; *Difference and Repetition* and the first Spinoza book, *Expressionism in Philosophy: Spinoza*, came out in 1968. However, Deleuze and Guattari began to correspond in spring 1969, then met during that summer. See Dosse (2007).
5. See for example Raymaker (1996) on Deleuze in light of Buddhist-Christian ethics.
6. In Danchin's translation of his 1998 *La Barque de Delphes* (*The Delphic Boat*, 2002) he significantly re-edits and revises the text, omitting some earlier material. Since Bifo cites from the French text, we necessarily consulted this text for the first citation.
7. Sarti is the coordinator of the research group in neuromathematics and visual cognition, and works both mathematics and in the department of Electronics, Informatics and Systems at the University of Bologna. See www.vision.unibo.it/ (accessed 30 August 2007).
8. We note that Kelly has two different books by the same title, *Out of Control* (Kelly 1994; 1995), neither of which Bifo includes in his own list of references (see Bibliography).

Chapter 11

1. We assume that Bifo's reference to 'Texas-style houses' and other features is meant to emphasize size, uniformity and ostentatious display of wealth.
2. See also the three aspects of the refrain in beautiful opening pages of Plateau 11, '1837: Of the Refrain' (*A Thousand Plateaus*, 311–12).

Interview

1. Armando Verdiglione (1944–), Italian psychoanalyst and semiotician. Verdiglione was originally trained as a literature scholar, then in the early 1970s he completed his psychoanalytic training in Paris with Lacan. Extremely compromised by his closeness to Italy's rich and famous, Verdiglione's scientific credibility greatly eroded in the 1980s and 1990s.
2. Franco Basaglia (1924–80), Italian psychiatrist and founder of the 'institutional psychiatry' movement in Italy. In the 1960s and 1970s, Basaglia was instrumental to a massive reform of psychiatric institutions in what was an extremely backward and repressive environment.
3. Guattari's essay is 'Les luttes de désir et la psychanalyse', republished in *La Révolution moléculaire* (1977, 29–43), and translated as 'Psychoanalysis and the Struggles of Desire' in *Molecular Revolution* (1984, 62–72).
4. Following the repression of the youth movement in Bologna and other cities in the summer of 1977, an anti-repression convention was organized for the days of 23, 24 and 25 September. Guattari was instrumental in its organization and participated personally in it.
5. According to François Dosse, Guattari and François Pain created Radio Libre Paris [Free Radio Paris] (subsequently called Radio Tomate), for which Bruno Guattari, Félix's son, took over the programming (Dosse, 2007, 360).
6. San Basilio is a sub-proletarian neighborhood in Rome, where in the late 1970s squatters gave life to highly visible political struggles.
7. Since the early 1960s, Bologna had a city administration mostly managed by members of the PCI, and the party was being accused of having become a locus of power.
8. For more detail on Telestreet, see Berardi (Bifo) et al. (2004).
9. The Centre-Left coalition won the elections in the spring of 2006. Since then, a few of the journalists and performers 'epurated' by Berlusconi have been called back.
10. These are the two names of a writers' collective composed of four or sometimes five members based in Bologna. They have published very successful historical-political 'narrative objects' – mid-way between a novel and a historical/biographical narration – among which we may cite *Q* (Einaudi, 2000) and *Asce di Guerra* (Einaudi, 2003).
11. These two essays are available in English in the volume *Time for Revolution* (Negri 2003).
12. See the work of the French economist Serge Latouche, who makes of '*décroissance*' – literally 'de- or un-growth', but in English the translation 'postdevelopment' has prevailed – the most important priority for affluent Western societies.
13. Marco Pannella (1930–), founder and historical leader of the Partito Radicale Italiano.

Bibliography

Works cited by Franco Berardi (Bifo), Gilles Deleuze, Félix Guattari

Berardi (Bifo), Franco (1974) *Scrittura et movimento* (Venice-Padova: Marsilio).

Berardi (Bifo), Franco (1980) 'Anatomy of Autonomy' *Autonomia. Semiotext(e)* 3: 3, 148–70.

Berardi (Bifo), Franco (1997) 'Corpo senza organi e divenire planetario nel pensiero Deleuze-Guattari', in Salvo Vaccaro (ed.) *Il secolo Deleuziano* (Milan: Mimesis Editore), 125–30.

Berardi (Bifo), Franco (2007) 'Schizo-Economy', translated by Michael Goddard, *SubStance* 36: 1, 76–85.

Berardi (Bifo), Franco, Alain Guillerme and Pierre Rival (1977) *L'Ideologia francese: contro i nuovi filosofi* (Milan: Squilibri).

Berardi (Bifo), Franco, Marco Jacquemet and Giancarlo Vitali (2004) *Telestreet: macchina immaginativa non-omologata* (Milano: Baldini e Castoldi).

Deleuze, Gilles (1983) *Nietzsche and Philosophy,* translated by Hugh Tomlinson (New York: Columbia University Press).

Deleuze, Gilles (1984) *Kant's Critical Philosophy,* translated by Hugh Tomlinson and Barbara Habberjam (Minneapolis: University of Minnesota Press).

Deleuze, Gilles (1988) *Bergsonism.* translated by Hugh Tomlinson and Barbara Habberjam (New York: Zone).

Deleuze, Gilles (1988) *Spinoza: Practical Philosophy,* translated by Robert Hurley (San Francisco: City Lights).

Deleuze, Gilles (1990) *Expressionism in Philosophy: Spinoza,* translated by Martin Joughin (New York: Zone).

Deleuze, Gilles (1990) *The Logic of Sense,* translated by Mark Lester with Charles Stivale; edited by Constantin V. Boundas (New York: Columbia University Press).

Deleuze, Gilles (1991) *Empiricism and Subjectivity: An Essay on Hume's Theory of Human Nature,* translated by Constantin V. Boundas (New York: Columbia University Press).

Deleuze, Gilles (1993) *The Fold. Leibniz and the Baroque,* translated by Tom Conley (Minneapolis: University of Minnesota Press).

Deleuze, Gilles (1994) *Difference and Repetition,* translated by Paul Patton (New York: Columbia University Press).

Deleuze, Gilles (1995) *Negotiations,* translated by Martin Joughin (New York: Columbia University Press).

Deleuze, Gilles (2000) *Proust and Signs,* translated by Richard Howard (New York: George Braziller (reprint) (Minneapolis: University of Minnesota Press).

Deleuze, Gilles (2006) *Two Regimes of Madness,* edited by David Lapoujade, translaed by Ames Hodges and Mike Taormina (New York: Semiotext(e)).

Deleuze, Gilles, and Félix Guattari (1983) *Anti-Oedipus. Capitalism and Schizophrenia*, 1. Trans. Robert Hurley, Mark Seem and Helen Lane (New York: Viking; Reprinted, Minneapolis: University of Minnesota Press).

Deleuze, Gilles, and Félix Guattari (1986) *Kafka. Toward a Minor Literature*, translated by Dana Polan (Minneapolis: University of Minnesota Press).

Deleuze, Gilles, and Félix Guattari (1987) *A Thousand Plateaus. Capitalism and Schizophrenia 2*, translated by Brian Massumi (Minneapolis: University of Minnesota Press).

Deleuze, Gilles, and Félix Guattari (1994) *What Is Philosophy?*, translated by Hugh Tomlinson and Graham Burchell (New York: Columbia University Press).

Deleuze, Gilles, and Leopold Von Sacher-Masoch (1989) *Masochism. Coldness and Cruelty*, translated by Jean McNeil (New York: Zone).

Guattari, Félix (1972) *Psychanalyse et transversalité* (Paris: François Maspero).

Guattari, Félix (1977) *La Révolution moléculaire* (Fontenay-sous-Bois: Recherches).

Guattari, Félix (1979) *L'Inconscient machinique* (Fontenay-sous-Bois: Recherches).

Guattari, Félix (1984) *Molecular Revolution. Psychiatry and Politics*, translated by Rosemary Sheed (New York: Penguin).

Guattari, Félix (1977) 'Des millions et millions d'Alice en puissance', introduction to *Radio Alice, radio libre*, translated by Danièle Guillerm and G. Marco Montesano (Paris: Laboratoire de Sociologie de la Connaissance, Université de Paris VIII).

Guattari, Félix (1984) 'Millions and Millions of Potential Alices', in *The Molecular Revolution*, 236-41. Translation of 'Des millions et millions d'Alice en puissance' (1977).

Guattari, Félix (1984) 'Plan for the Planet', in *The Molecular Revolution*, 262–72.

Guattari, Félix (1986) *Les Années d'hiver, 1980-1985* (Paris: Bernard Barrault).

Guattari, Félix (1986) 'Gérard Fromanger, la nuit, le jour' [1984], in *Les Années d'hiver*, 248-58.

Guattari, Félix (1995) *Chaosmosis, an Ethico-Aesthetic Paradigm*, translated by Paul Bains and Julian Pefanis (Bloomington: Indiana University Press).

Guattari, Félix (1997) *Piano sul pianeta. Capitale mondiale integrato e globalizzazione* (Verona: Ombre corte).

Guattari, Félix (1999) 'Jean-Jacques Lebel: Painter of Transversality' [1988], translated by Melissa McMahon, at: www.artdes.monash.edu.au/non-cms/globe/issue8/jjltxt.html (accessed 19 June 2007).

Guattari, Félix (2000) *The Three Ecologies*, translated by Ian Pindar and Paul Sutton (London: Athlone). Translation of *Les trois ecologies* (Paris: Galilée, 1989).

Guattari, Félix (2002) *La philosophie est essentielle à l'existence humaine, entretien avec Antoine Spire* (Paris: Editions de l'Aube).

Guattari, Félix (2006) *The Anti-Oedipus Papers*, edited by Stéphane Nadaud; translated by Kélina Gotman (New York: Semiotext(e)).

Guattari, Félix, and Eric Alliez (1984) 'Capitalistic Systems, Structures and Processes', translated by Brian Darling. In *Molecular Revolution*, 273–87. Translation of 'Le capital en fin de compte: Systèmes, structures et processus capitalistiques', *Change international* 1, 1983, 100–6.

Guattari, Félix, and Antonio Negri. (1990) *Communists Like Us,* translated by Michael Ryan (New York: Semiotext(e)). Translation of *Les Nouveaux espaces de liberté* (Paris: Dominique Bedou, 1985).

Other works cited

Adorno, Theodor W. (1974) *Minima Moralia: Reflections from Damaged Life,* translated by E. F. N. Jephcott (London: NLB).

Adorno, Theodor W. (1979) *Negative Dialectics,* translated by E. B. Ashton (New York: Seabury Press).

Adorno, Theodor W. (1993) *Hegel: Three Studies,* translated by Shierry Weber Nicholsen (Cambridge, MA: MIT Press).

Affinity Project (2007) 'Autonomist Marxism', at: www.affinityproject.org/traditions/autonomousmarxism.html (accessed 7 June 2008).

Balestrini, Nanni (1971; 2001) *Vogliamo tutto* (Milan: Feltrinelli; Rome: Derive Approdi).

Barbrook, Richard (2003) *The Holy Fools,* at: http://libcom.org/library/deleuzo-guattariansbarbrook (accessed 16 February 2007).

Bataille, Georges (1988) *Inner Experience,* translated by Leslie Anne Boldt (Albany: State University of New York Press).

Bateson, Gregory (1979) *Mind and Nature: A Necessary Unity* (New York: E. P. Dutton).

Bergson, Henri (1965) *Duration and Simultaneity,* translated by Leon Jacobson (Indianapolis: Bobbs-Merrill).

Bergson, Henri (1983 [1911]) *Creative Evolution,* translated by Arthur Mitchell (Lanham: University Press of America).

Bista, Dor Bahadur (1991) *Fatalism and Development. Nepal's Struggle for Modernization* (Calcutta: Orient Longman Ltd).

Broeckman, Andreas *Minor Media – Heterogenic Machines,* at: http://isp2.srv.v2.nl/~andreas/texts/1998/minormedia-en.html (accessed 30 August 2007).

Burroughs, William (1982) *Ah Pook Is Here* (New York: Riverrun Press).

Collective A/Traverso (1980) 'Radio Alice – Free Radio' *Autonomia. Semiotext(e)* 3: 3, 130-4.

Danchin, Antoine (1998) *La Barque de Delphes* (Paris: Editions Odile Jacob).

Danchin, Antoine (2002) *The Delphic Boat. What Genomes Tell Us,* translated by Alison Quayle (Cambridge, MA: Harvard University Press).

Derrida, Jacques (1978) *Writing and Difference,* translated by Alan Bass (Chicago: University of Chicago Press).

Descombes, Vincent (1980) *Modern French Philosophy,* translated by L. Scott-Fox and J. M. Harding (Cambridge and New York: Cambridge University Press).

Dick, Phillip K. (1968) *Do Androids Dream of Electric Sheep?* (New York: Doubleday).

Dosse, François (2007) *Gilles Deleuze, Félix Guattari. Biographie croisée* (Paris: La Découverte).

Fabbri, Paolo (1997) 'Come Deleuze ci fa segno', in Salvo Vaccaro (ed.) *Il secolo Deleuziano* (Milan: Mimesis Editore), 111-23.

Fuenmayor, Ramsés (1991) 'Truth and Openness: An Epistemology for Interpretive Systemology.' *Systems Practice* 4: 5, 1-17, at: www.saber.ula.ve/db/ssaber/Edocs/centros_investigacion/csi/ publicaciones/papers/fuenmayor-truth.pdf (accessed 18 June 2007).

Genosko, Gary (2002) *Félix Guattari: An Aberrant Introduction* (New York: Continuum).
Genosko, Gary (2003) *The Party Without Bosses. Lessons on Anti-Capitalism from Félix Guattari and Luís Inácio 'Lula' da Silva* (Winnipeg: Arbeiter Ring Publishing).
George, Susan (1988) *A Fate Worse Than Debt* (New York: Penguin).
George, Susan (1992) 'La dette se paie en nature', *Le Monde diplomatique*, supplement 'Le Sommet de Rio-di-Janeiro', June, pp. ii, vi.
George, Susan (1992) *The Debt Boomerang* (London: Pluto Press).
Hagège, Claude (1990) *The Dialogic Species*, translated by Sharon L. Shelly (New York: Columbia University Press).
Hardt, Michael (1993) *Gilles Deleuze. An Apprenticeship in Philosophy* (Minneapolis: University of Minnesota Press).
Hardt, Michael, and Antonio Negri (2000) *Empire* (Cambridge, MA: Harvard University Press).
Hegel, G. W. F. (1977) *The Phenomenology of Mind*, translated by J. B. Baillie (Atlantic Highlands: Humanities Press Inc).
Heidegger, Martin. (2000) *Off the Beaten Track*, translated by Julian Young and Kenneth Haynes (Cambridge: Cambridge University Press).
Hume, David (1888) *A Treatise of Human Nature*, edited by L.A. Selby-Bigge (Oxford: Clarendon Press).
Jung, C. G. (1963) *Mysterium Coniunctionis*, translated by R. F. C. Hull (London: Routledge and Kegan Paul).
Kelly, Kevin (1994) *Out of Control. The Rise of Neo-biological Civilization* (Reading, MA: Addison-Wesley).
Kelly, Kevin (1995) *Out of Control. The New Biology of Machines, Social Systems, and the Economic World* (Reading, MA: Addison-Wesley).
Marcuse, Herbert (1941) *Ragione e rivoluzione. Hegel e il sorgere della 'teoria sociale'* (Bologna: Il Mulino Biblioteca).
Marcuse, Herbert (1997) *Reason and Revolution. Hegel and the Rise of Social Theory* (London: Routledge and Kegan Paul).
Marx, Karl, and Friedrich Engels (1972) *Die heilige Familie* (Glashütten in Taunus: D. Auvermann).
Mayakovsky, Vladimir (1986) 'Longer Poems'. in Volume 2 of *Vladimir Mayakovsky. Selected Works in Three Volumes*, translated by Dorian Rottenberg (Moscow: Raduga Publishders).
Negri, Antonio (2003) *Time for Revolution*, translated by Matteo Mandarini (London: Continuum).
Nietzsche, Friedrich. (1935, 1937) *La Volonté de puissance*, translated by G. Bianquis; edited by F. Würzbach (Paris: NRF).
Nietzsche, Friedrich (1968) *The Will to Power*, translated by Walter Kaufmann and R. J. Hollingdale (New York: Random House, Inc).
Nietzsche, Friedrich (1983) *Untimely Meditations*, translated by R. J. Hollingdale (Cambridge and New York: Cambridge University Press).
Nietzsche, Friedrich (1996) 'The Wanderer and His Shadow', in *Human, All Too Human*, translated by R. J. Hollingdale (Cambridge and New York: Cambridge University Press).
Prince, Bernard, and François Videcoq (2006) 'Félix Guattari et les agencements post-media, at: http://multitudes.samizdat.net/article1974.html (accessed 26 August 2007).

Proust, Marcel (1993) *Time Regained*, translated by Andreas Mayor and Terence Kilmartin; revised by D. J. Enright (New York: The Modern Library).
Raymaker, John (2006) *Empowering Philosophy and Science with the Art of Love: Lonergan and Deleuze in the Light of Buddhist-Christian Ethics* (Washington, DC: University Press of America).
Spinoza, Baruch (1992) *The Ethics*, translated by Samuel Shirley (Indianapolis: Hackett Publishing).
Trongpa, Chogiam (1986) *Méditation et action* (Paris: Seuil).
Vaccaro, Silvio (1997) *Il secolo Deleuziano* (Milano: Mimesis Editore).
Vattimo, Gianni (1993) *The Adventure of Difference*, translated by Cyprian Blamires (Cambridge: Polity Press).

Index

position of absolute disaffection with politics, and so this probably played a role, but I considered Félix's activism to be an obstacle, an excess of generosity in a situation that did not deserve it and from which he had little to gain.

GM: *That is really very interesting. I thought that basically this struggle of the Greens did not entirely remain a dead letter, that is, do you really think that their work did not amount to anything? For example, the entire politics against the GMOs ...*

Bifo: No, no, and I can even say more, that is, we can say that the Greens' thinking and their practice in the last twenty years determined the modification of what you were discussing quite well. That is, it became possible to speak of globalization from a critical point of view. The Greens' work made us understand that processes don't only have a local or national dimension: this is all quite good. I am far from disputing this, but I found unproductive the Greens' political commitment in the 1980s. On the other hand, I believe myself to have been sufficiently commited, in the last five years, in the so-called 'non-global' movement! In fact, I would have been very happy if Félix had still been with us in those years because I am certain that he would have been able to make an extraordinary contribution, one that would have been truly recognized!

But why? Why, in fact, was the Greens' practice in the 1980s ... I know the Green world very well, quite sufficiently from inside. My sister became very involved in those events, they are my friends, I voted with them, I hung out with them, etc., but what was the Green world in the 1980s? There were thousands of extraordinary practices: in therapy, in experimentation on alternative energies, etc. So thousands of extremely interesting daily practices, and on the other hand a political manoeuvring of the lowest possible order! That is, they never succeeded, but perhaps it was just as well this way ... The Green movement is not a political movement, but rather a movement for reorienting politics. But when it gets reduced to participation in elections, then perhaps it is not that good, with the possible exception of the Germans. To sum it all up, I would say that the parliamentary engagement of the Italian Greens does not reflect at all the importance of their conceptual contributions. And so I told Félix, 'But why should you get involved in all that?'.

GM: *Write books!*